Eyes Open
Looking for the Twelve

Blueprint for a New World

By

Jeanne White Eagle

Eyes Open – Looking for the Twelve

Copyright © 2014 by Jeanne White Eagle

Manufactured in the United States of America. All rights reserved. No other part of this book may be reproduced in any form or by any electronic or mechanical means including information storage and retrieval systems without permission in writing from the publisher, except by a reviewer, who may quote brief passages in a review. First edition.

ISBN: 978-1497413085 (paperback)

Cover art and design by Sean Pehrson

Illustrations by Jeanne White Eagle

EYES OPEN

Looking for the Twelve

Jeanne White Eagle

Acclaim for Jeanne White Eagle's
Eyes Open – Looking for the Twelve

"In *Eyes Open*, author Jeanne White Eagle depicts a vision for a world that works for everyone, a world where transformation, an enlightened population and paradigm-popping realities are the norm rather than the exception. Her story may challenge some of your assumptions, but then aren't assumptions something we should challenge every once in a while anyway? After all, hasn't most of our progress as a species been largely due to somebody challenging some commonly-held assumption, like humans can never fly, or land on the moon, or….well, we you get my drift."

- John Renesch, futurist, author, *Getting to the Better Future* and *The Great Growing Up*

<>

"*Eyes Open* offers readers a wide-ranging vision for the future of humanity ‹ a future with a radically different form of education and a central role for imagination, a future in which we respond to change with love, not fear, recognizing the vibrational basis of reality and the importance of 'gardening our hearts.' By 'slipping between the slices of light' to get clear of ego, in this new future we become part of a "spacefaring community" that previously lived only in the worlds of *Star Trek* and other sci-fi works. But *Eyes Open* is not science fiction. It is, rather, the rarest of the rare ‹ a book based on dreams, fleshed out in great detail, carried forward with vivid imagination and vision, offering the reader a call. Are you one of the 12? Reading this book, taking in its message could be the most momentous, life-changing event of your life!"

- Susan Mehrtens, Ph.D., President,
  The Jungian Center for the Spiritual Sciences

<>

"*Eyes Open* opens us to the possibility of a radically new way of learning and communicating: beyond the boundaries set by race, religion, political persuasion, nation state or even planet of origin. The author, Jeanne White Eagle, extends an invitation for the reader to join her in manifesting a shared dream of creating the physical, environmental and systems structures needed to enable and nurture an enlightened international and interplanetary community."

- Herman Maynard, business executive and co-author of
  The Fourth Wave: Business in the 21st Century

Dedicated to

My beloved husband, John Pehrson who stands by my side,

Our amazing children, Jenny Stiffler and John Troutner,

Ryan, Sean and Alan Pehrson

My Mom and Dad, Kit and Perry,

who helped me to believe in my dreams

and

My precious and beautiful granddaughter Kendolynn

and all other Star Children

who have come here now to inspire and guide us.

Contents   Page

Foreword:
by Astronaut Edgar Mitchell,
NASA Moon Mission, Apollo 14

BEGINNING

Prelude to a Dream: Eyes Open 1

PART 1: WAKING UP

Chapter 1: Background 13

Chapter 2: A Foretelling 17

Chapter 3: What's In a Name 23

PART 2: THE DREAM

Chapter 4: The Dream 27

Chapter 5: A Map 31

Chapter 6: Two Parts of a Whole 35

Chapter 7: Five Circles 39

Chapter 8: Willis Talks 43

Chapter 9: Significance of Thirteen *51*

Chapter 10: Shape and Orientation *55*

Chapter 11: The ICC (Interstellar Communication Center) *61*

Chapter 12: "Universe Cities" *77*

Chapter 13: Pyramids within Pyramids *81*

Chapter 14: Sound *85*

Chapter 15: Crystals *95*

PART 3: THE TRANSMISSIONS

Chapter 16: Prelude *101*

Chapter 17: Transmissions *105*

Chapter 18: Second Recurring Dream: A Transit System *147*

PART 4: FORWARD

Chapter 19: Love *153*

Chapter 20: Paradigm Shift *161*

ILLUSTRATIONS

 # 1: Sketch of Recurring Dream: Scene 1 .... *169*

 # 2: Sketch of Recurring Dream: Scene 2 .... *170*

#3: 13 Circles (identified) of U/ICC *171*

#4: Golden Pyramids *172*

#5: 13 Circles *172*

#6: Energy Pyramids on 13 Circles *172*

#7: Energy Beacons(Signals) resulting from Pyramid Energy flow in U/ICC *173*

 # 8: Two-dimensional Bipolar Magnetic Field (Prana Tube) created by the structural design and energy of U/ICC *174*

<>

Acknowledgements *177*

Books of Reference and Excellent Reads *183*

About the Author *189*

Index *191*

FOREWORD

Because of the progress in science and technology during the 20th and early 21st century the population of our planet has virtually exploded from less than two billion persons in 1900 to over 7 billion persons in the early 21st century. Although there are many wonderful benefits for humankind and animal kind from science and technology, the downside is increased consumption and loss of non-renewable resources needed for survival and a global warming that is altering the natural processes that nurtured the planet and its species over the centuries. The likelihood is that these excesses threaten the very survival of life on Earth unless we humans can find the means to offset them.

Fortunately, understanding of the role of consciousness in humans and in nature in general, has been studied for millennia, and during the most recent times as an appropriate topic in science. Learning to serve the greater good of all living things has been a noble practice taught by ancient seers and mystics. The modern studies are learning to draw on this ancient wisdom and offer the hope that we can bring together the powerful technologies of the modern period and apply them to the serious condition of our modern civilization that is approaching a point of no return due to the immense success of technological production.

The power of love that allows us to reach out to each other and to nature itself has been taught by mystics, seers and wise persons for millennia. It brings the healing and the satisfaction that we all need in order to function well on this little planet. We will be wise if we apply these lessons to our contemporary dilemmas.

Jeanne White Eagle in *Eyes Open* discusses these vital issues from the point of view of ancient wisdom traditions as well as utilizing the modern studies from science in the studies of consciousness.

Though many recognize and have voiced the existing challenges that face our world, and even though there have been many technological advances made to offer clues to the survival of Earth, *Eyes Open* actually offers a *possible "how-to" blueprint* to put us on the track toward achieving a higher stage of consciousness and evolution, ensuring the continued existence of life on Earth.

Reading more like a science fiction, *Eyes Open* rings of a mission to find those out there who understand the dilemmas our world now faces and who are called to be part of the adventure to resolving those dilemmas. This includes moving us forward into a future that embraces a more evolved human species capable of healthy interaction with itself and the rest of life on our planet, and in time, with life in the larger Universe.

>   - Edgar Mitchell, Astronaut
>   Apollo 14 NASA Moon Mission,
>   Founder of Institute of Noetic Sciences
>   (IONS)

BEGINNING

PRELUDE TO A DREAM

"Eyes Open"

There's a small book out there that I read many years ago. It's called *ET 101, The Cosmic Instruction Manual for Planetary Evolution.*[1]

I remember the impact this curious little book had on my husband and me. We began reading it for the first time *out loud* while driving through the countryside of southwestern Colorado. We were struck by its simplicity and humorous way of speaking about topics of a dire nature…namely the survival of Earth.

At first, neither of us got it that this pocketsized manuscript was anything but a colorful and inspired idea of some creative author with a lively imagination who had a talent for writing in a style that most people could hear.

So, who was the author?? We looked everywhere through the book's pages to no avail. The only thing we did find was a very tiny reference located in the bottom of the very back page of the book that said: "Co-Created by Mission Control and Zoev Jho." Then there was a dedication

---

[1] *ET 101*, published by HarperCollins, can be found through Amazon.com and other online booksellers.

from the "Intergalactic Council." The letter-size of this information was so diminutive we almost overlooked it.

We continued to read on in this small jewel, enjoying the humor and wild ideas it proposed, unaware at first, of the chord of truth the book was touching. It didn't take long, however, before it dawned on both of us that we were reading something quite profound. This happened less than half way through the book. From that point on we began to read in earnest, conscious of what might lay between the lines.

What we learned inspired us and managed to make its mark on the years that would follow. There is no doubt that the influence this tiny book had on each of us has lasted to this day. I share this short story with you because …

As you read through *Eyes Open*, you might want to keep your "eyes open" to read *between the lines*, just as we did with *ET 101*. If you find yourself at the precipice of such thoughts as, "This can't be real," or "Is she serious?" don't be surprised if you discover that the answers are, "Yes, this is real." "Yes, she *is* serious."

I would like to think, however, that there are bits of humor scattered here and there for spice and easier digestion for you, the reader.

So let's get on with it…

<>

I have never seen myself as a writer. But, I, as with most people, am someone who can be driven to do the near impossible when a principle of *truth* is at stake and the cause is a noble one.

In this case, the "near impossible" is my writing this book with the hope of successfully communicating an idea of such immense magnitude that it too tastes of the *impossible*.

It all stems from a *dream, a recurring dream* had again and again over a period of more than two decades. It's a dream that proposes a possibility for the higher development of humanity, its role in the survival and evolution of all life on Earth, and the eventual open communication and interaction with life beyond our planet.

'Sounds a bit grand from the onset, doesn't it? But consider this. It's time.

Like it or not, we are finally coming face to face with the results of what we as a species have done to ourselves and our planet.

For example, it was only a short while ago that we surpassed the number of seven billion people on Earth, barely conscious of the fact that to feed and nourish this many human beings without abusing our home world is near to impossible. Present day solutions would most likely result in a near-to-total stripping away of the natural resources that are already becoming less and less abundant. Feeding an overpopulated world of this magnitude would demand a level of maturity and awareness that humankind has only now begun to address. And if the United Nations Food and Agriculture Organization is correct, between the years of 2010 and 2012 one in eight people worldwide was chronically malnourished. If you do the math, that is over 870 million people starving on planet Earth.

With overpopulation have come diseases and viruses that are spanning the globe at an alarming rate, requiring wisdom and cures that, in too many cases, appear to be outside of present day knowledge and understanding. On top of this, whether out of ignorance and/or the need to *control* our world, humanity has managed to pollute Earth's air, land and water ... in the name of "progress." This includes the destruction of Earth's forests whose existence plays a fundamental role in producing the very air we breathe.

These are only a few of the self-made challenges that humankind now faces, all of which brings us to yet another quandary: How can humans care for their home planet when so much effort is spent on annihilating each other, creating nuclear and biological weapons that ultimately will end up destroying themselves as well as all other life on Earth? ...

Credo Mutwa[2], with whom my husband and I had the great honor of spending many hours during our years visiting

---

[2] Credo Mutwa is considered by many to be the "Father of Africa." He is a Sangoma (traditional African healer) that has the honor of being called Vusamazulu (Awakener of the Zulu) and Sanusi (The Uplifter). Credo is the keeper of the ancient wisdom

South Africa, has made the comment that human beings have turned into a virus-like animal on Earth, inflicting changes on a world that "is not ours." To quote him, "The human race is destroying Earth. Why? We know it is wrong to cut the forest in the jungle. We know it is wrong to spill oil into the sea. We know it is wrong to make many chimneys and pour smoke into the air. But we are doing it."

The mindset of conscious manipulation of life's building blocks that is inadvertently resulting in the annihilation of many of Earth's lifeforms and threatening the existence of the remaining living species on Earth, is not okay. And now an alarm bell that began ringing some time ago has reached its peak. The time of choice is here. Do we humans continue down the road to our own demise OR do we choose to change our minds, change our ways and change the present reality. Impossible???

Willis Harman, in his book *Global Mind Change,* says, "By deliberately changing the internal image of reality, people can change the world. Perhaps the only limits to the human mind are those we believe in." It is time to believe differently, to change our minds and create a world that richly nurtures and supports all life.

It is also time to allow our creative juices to flow. To reach far beyond ourselves into a Universe waiting with open

---

and relics of his people and is considered to be a prophet and great teacher. We met him years ago during our travels to South Africa and spent many wonderful hours partaking of his stories and wisdom. Some of our experiences with Credo are included in our biography, *Journey for the One,* by Monty Joynes (2008).

arms to receive a young species as ours full of curiosity and potential.

Enter Gene Roddenberry. Like so many of you, for decades I have been influenced profoundly and passionately by the genius of Roddenberry and his creation of Star Trek. There is the famous quote that has inspired generations and has managed to move us toward a future that is already showing itself in new and ever increasing breakthroughs affecting countless facets of our daily life:

> "To explore strange new worlds;
> to seek out new lifeforms and
> new civilizations;
> To boldly go where no one has gone before."

There will be those who say that what I have seen cannot be done. There were also some who said we'd never land on the moon. The way I think it happens is this: *If you can see it, you can do it. Anything is possible. Anything.*

I believe that what I have seen in my *recurring dream* is not only possible, it is necessary. My hope is that it will help to inspire humanity to reach out and grasp the future with boldness, confidence and courage. This book is one of the foundation stones being put into place to bring it all about.

So how did I arrive at the title *Eyes Open?*

It happened in the blink of an eye. I was visiting my daughter and was watching an episode of Star Trek Voyager. After several moments into the story, I heard the phrase "Eyes Open." When it was repeated again and then again, I suddenly realized that what I was hearing was the title of my book. This was something I had been searching for, for months, knowing I would recognize it when I found it.

This particular Voyager episode was one in which a reptilian alien paleontologist on a distant planet traced the roots of his species back millions of years. He discovered that the origin of his people came from planet Earth. And with great surprise, he found a common ancestral DNA link between his own cold-blooded species to that of warm-blooded humans.

As improbable and unbelievable as this finding was, this alien scientist's constant thought was *"Eyes Open."* This was his inspirational goad to remind himself that however impossible something may seem, it is not only necessary but crucial to keep one's self open to the possibility of the *impossible* being true. And as a responsible member of an intelligent species, it was imperative that one always be willing to explore beyond known limits however convinced an individual or group may be that something is factual and that the process of its discovery is infallible.

Unfortunately, when presented with unquestionable proof, the leaders of the alien civilization were threatened by the scientist's findings, rejecting his theory with zealous anger. To have acknowledged the new-found conclusions would have changed their species' entire belief system around their origin and upset the foundation of their whole society.

My husband John and I have a Native American friend who speaks of the trap created when one gets "stuck in form," suggesting that traditional ways of thinking and being can often be a trap that prohibits journeying into the unknown. Another way to say this is getting "stuck in form" can keep *new ways of thinking and being* from entering through the doorway to unlimited possibilities.

Pioneering Harvard Professor, Dr. John Mack, in his book *Passport to the Cosmos,* speaks of the phenomenon of alien abduction and the different viewpoints coming from a wide variety of sources and perspectives. Whatever a person's field of training, expertise or interest, be you a scientist or farmhand, Mack suggests that each of us may be being "invited to change our ways, to expand our consciousness and ways of learning, to use, in addition to our conventional ways of knowing and observing, methodologies more appropriate to… [the] complex, subtle and perhaps ultimately unknowable nature of life outside the box of what we humans think we know and believe we can *prove* without a doubt." Mack goes on to make a plea that we "begin with an attitude of *not* knowing … that we suspend the natural tendency to form questions according to the logic that is characteristic of the human mind as it applies itself to the human world." This is what I'm asking of you, the reader.

Travel with me now into a world where the *impossible* becomes a welcomed and familiar doorway into a new future. Give yourself the go-ahead to soar the heights with me, to push the envelope of what-has-been into what-can-be.

PART I

Waking Up

" *The Wright Brothers couldn't invent the airplane*
without having faith that humans could fly.
To stop war, we must first imagine peace.
To eliminate poverty, we must imagine abundance.
To paint a masterpiece, we must imagine beauty.
To change, to make a fresh start, and
To live up to our highest fullest potential,
We must not only imagine
a better future for ourselves,
But imagine it over and over
at each step of the way,
Knowing that the sun still exists
Even when it is night.

- From Naomi Hoshimo Horii's
"The Mind's Eye"

CHAPTER 1

Background

There are twelve of you out there. I'm looking for you. Twelve who have had the same *dream* I have had.

Perhaps yours was a *recurring dream* too. Mine started in my late teens. It continued into my thirties when it abruptly stopped only to suddenly reappear in 2009, more than twenty-five years later. *As I write this I find my thoughts are moving at lightning speed, as though some part of me is rushing to make up for all the years that have passed since I first had the dream.*

Not wanting to get ahead of myself, maybe in this moment it's important only that you know why I'm writing this book. It is to find you. Whoever you are. Wherever you are.

<>

When the *dream* returned in 2009 I had to take some deep breaths and step back to get perspective as I found myself questioning, with some scrutiny, the scope of what I'd seen. I believe I was being shown a way to ensure that our human species evolves in such a manner that all life on our planet not only survives but does so *in community*. Healthy. Vibrant. Caring. Sustainable.

As if this weren't enough, there was more. In fact, the *more* is the stuff of which science fiction is made ... *fiction* that my *dream* was telling me will one day become *fact*.

Now anyone who knows me knows I don't shy away from looking at the big picture of anything. Example. For years I worked in my dad's company, an international business that made and sold training videos to corporations around the world. I quickly discovered that it took no more energy and creativity to sell the entire lot of our programs in one fell swoop than it did to sell a single program. It simply required the willingness to see from a *larger* perspective ... i.e. to *think outside the box*.

There was the time I was ten years old, home sick with the measles. I was watching a television show advertising a competition that awarded one hundred dollars to whoever the hundredth caller was. I remember going through the thought process of weighing the odds in my young mind, coming to the conclusion that even though I was a child, I had as good a chance as any adult to win the contest. I made the call and won the prize. An interesting footnote is that I asked my mom to gift part of my hundred dollars to a certain non-profit organization I had also seen advertised on TV that helped to feed and take care of children in the world who had no homes. My thinking then, though perhaps not in these terms, was knowing that regardless of my being a young child I still might be able to make a difference.

I remember being sixteen and in my college dorm reading a Sunday newspaper. I read about a Chinese man who had defected from his country during a time of great upheaval and was trying to get his wife and children, who were still in China, out of the country safely. I was so touched by his story that I decided to write to people all over the world asking them to pray and send good thoughts to this man and his family for their safe reunion. Again, even at a young age, my thought process was one of a global scale. It made sense to me that invoking many people in this effort would carry a lot more weight in the ethers than my single individual effort. So I wrote to various friends and relatives. Plus, having attained addresses, I wrote to leaders of countries and influential people around the world whom I did not know but thought would want to help. This included President and Mrs. Eisenhower, who graciously wrote a kind letter of positive response some days later.

I have many stories including experiences that occurred in my travels with the international musical Up With People, my time with the World Business Academy and more recent ones involving the following of a global vision I received some years ago of a ceremonial dance of peace that has now gone worldwide. The point of my sharing any of these stories is to say I'm no stranger to "thinking big."

Yet, when I saw the scope of what my recurring *dream* was showing me, I have to admit that in addition to the elation and hope I was seeing for our future, I found there were butterflies flying around inside my belly that hadn't yet found formation. Why? Why, when I would think about the *dream* I immediately got the same sensation one experiences at the top of a roller coaster where the tummy seems to do a flip flop of sorts at the highest point and the awareness of the unknown kicks in? Was it because I was also seeing what my role in all this was to be…which is to move this vision through the birthing canal? And to find those of you who will bring the *dream* I was given into its full physical manifestation?? The questions kept coming, "How am I to do this? Is this something I am capable of?"

When the truth of all this began to register with me, I experienced something akin to *stage fright*. I had to keep reminding myself to breathe. I was determined to help my belly butterflies find a pattern of flight that would bring clarity and focus.

CHAPTER 2

A Foretelling

It was in the late 1980s that I received a call from my friend, Sue Mehrtens.

Sue is probably one of the most fascinating and brilliant people I know. Intellectually, she is a strong practical thinker. She speaks several languages, can read a thousand pages in one day, and has periodically served as a consultant for such global corporations as IBM and Hewlett Packard. At the same time she has the unique ability to move her ego out of the way with a childlike innocence that allows something larger to come through, tapping into what many might call the *great mystery*. Among her extraordinary abilities, Sue is a genuine world class *seer*. She is someone that in the past might have been called a *prophet,* having the capability to erase timelines and see the past, present and the future all in the same sphere of reality. She walks with ease on the line between linear and non-linear phenomenon. A truly fascinating person!

Nothing could have surprised me more than to hear Sue's excited voice through the phone saying she was going to come to me at once! This meant driving from Maryland where she was living at the time to my home in Tennessee.

There was an urgency, a resolute sound in her voice as she made the statement that she was coming to visit for three days. Her call was prompted by the results of an astrological reading that I had requested from her a few days before.

Looking back on that fateful visit I can truthfully say it marked a turning point in my life. What Sue shared with me seemed to act as a trigger to awaken something lying dormant within me, something of profound significance that had been with me since birth. I just didn't know this at the time. Not until it happened. Here is what she told me…

Given all the elements in my astrological chart, apparently I was born with seven *Yods* and three *Grand Trines*. When she first told me, I looked at her with a blank face. I had absolutely no idea what she was talking about. She explained. A *Yod,* also called the *Finger of Fate* or the *Finger of God,* indicates a special life purpose. It creates the sensation of steadiness, special passion, deep aspiration, and a strong inner drive to accomplish something very important and special. It is considered a *point of destiny.*

She went on to tell me that the *Grand Trines* were particularly significant because this meant I had been given everything necessary to help me complete or fulfill my destiny.

Sue shared that it is rare for a person to have even one *Yod* so when she discovered that I had *seven(!)* she saw this as important information, enough to warrant her immediate visit.

One would have to admit this was pretty heady information and regardless of some of the less attractive aspects I have since learned about *Yods,* what I learned over those three days had a startling effect on me. At first my ego was challenged as I found myself experiencing what one might encounter having been presented some grand award. But when Sue got to the core of the information she had come to share, in those exact moments there were giant thunder claps from outside the window, followed by bright dramatic lashes of lightning! (No kidding! I felt like I was in a movie!) In an instant any puffed-up-ness the ego enjoyed was short lived as the thunder sounded and Sue said, "Jeanne, with this comes *great* responsibility." The stark realization of what this meant began to rush in with crystal clear awareness.

My belly butterflies were going berserk trying to make sense of it all. Yet the odd thing is that I could feel an ancient yawn taking place within me. Truly something was waking up. I was beginning to remember.

<>

Over the next years I found myself becoming immersed in daily living and the dramas that often entails. As a result I put the memory of Sue's visit and the impact of the information she shared into a mental file that only rarely did I retrieve and reflect upon.

In retrospect, though I saw the vast possibilities inherent in my dream and though Sue's information confirmed what at some deep level I already knew about myself, I was reluctant to share what I had learned with anyone. I kept asking myself, "What good would it do? Who in their right (sane?) mind would believe me? Who could begin to hear what I know without first thinking: 'Who does she think she is?' or 'How dare her think in such grandiose terms,' or 'Is she putting herself above me, above us?' etc. etc. etc." And so, I stayed silent. "Mum" became the pill I would swallow daily until it simply became a habit. Once a habit, all went back into the file. "Out of sight, out of mind." Almost.

<>

It wasn't until many years later when the *recurring dream* returned that I felt the old familiar nudgings set into motion from Sue's visit in the eighties. At the risk of sounding melodramatic, I knew my time had come.

Even if I had wanted to, there was no way to duck or hide. I had been given a job and I had said "yes." Probably long before I was ever born.

At the time, I was involved in carrying out another vision I'd been given of a singing ceremonial dance to help transmute the perception of *separation* in our global consciousness. A peace dance. As I look back, I see that this experience was a last effort on Life's part in preparing me to step forward to follow, once and for all, a calling set forth in the *dream*. Its voice was coming through *loud and clear*. The directive was unquestionable. I knew then as I know now that the path I would travel, would consume my time and energy for the rest of my years here on Earth. The final outcome will be my legacy.

<><><>

CHAPTER 3

What's in a name?

If you've grown up and been conditioned in the Western culture, particularly in the United States, then the very name *Jeanne White Eagle* immediately brings to mind for some a couple of possibilities. One and the most obvious is, "Oh, she's Native American." The answer? Yes, to a point. But not a full-blood. Cherokee on my Mom's side. I also have Irish, Scottish, English, French and Nordic traces roaming around in my bloodline somewhere. As with so many of us, I'm a two-legged melting pot.

Another possibility, especially for any who are familiar with my background either in the "growing up" years, in college and the immediate years that followed, or more recently in my corporate life in the business world, the name *"…White Eagle"* may expose certain prejudices. This is particularly true for all who have known me as *"Jeanne + single/married name"* but certainly not *"White Eagle"* tagged onto *Jeanne*. It doesn't seem to matter that the name of White Eagle was given to me by a highly respected Medicine Man and Mystic of two Native tribes in the American Southwest.

So why am I telling you this? And what does this have to do with my *recurring dream*?

… It is because this may be yet another way to glean the audience of those who will read this book. Any of the Twelve I am seeking will naturally see beyond the obvious and will resonate with who I am and why I write this book. Any of the rest of you who resonate with what follows in this book, you're invited to consider yourselves on board and part of one of the most amazing adventures yet imagined. And, indeed, if you choose, you are helping to midwife what is very possibly the greatest purpose for which any of us have been born.

And so…

PART II

The Dream

"You never change things by fighting the existing reality. To change something build a new model that makes the Existing model obsolete."
- Buckminster Fuller

CHAPTER 4

The Dream

Here is what I remember.

The first time the dream came I must have been around sixteen years old.

> *I'm standing at the foot of a very high hill or small mountain that is robed in green grass. No trees that I can see. To my left is a sea or very large lake. I'm looking up to see what looks like a complex of futuristic buildings, circular in shape. There look to be five large ones and several smaller ones within the complex.*
>
> *It seems as though the sun is in the early stage of rising or final phase of setting. I'm not sure which, one being as dramatic and beautiful as the other.*
>
> *The reflection of the light on the buildings, which appear to be white in color, creates an effect not unlike the phenomenon that occurs when light meets the face of a diamond. A brilliance happens that fills up the space of one's attention making it difficult not to notice that something extraordinary is beckoning.*

Then the scene changes.

Now I'm standing inside one of the buildings and I'm aware of people moving and interacting with each other and their surroundings as though they are in a training facility or school of some sort. But something is different. Strange at first. Yet there's a natural, even a congenial feel to it all…

As I allow the scene to come into full focus, I see faces of beings that I'm aware are not from our planet Earth. The "people" moving through the hallways, some stopping to engage in conversation, are a mixture of human beings and a variety of interstellar visitors.

Here my dream ends. It returned time and time again, until I was well into my thirties. Then, it stopped.

Its return close to thirty years later couldn't have surprised me more than it did. It made a grand appearance, getting my attention with cosmic pomp that would rival anything Hollywood could imagine. It came with technicolor clarity and new information. I was being invited to embark on a journey of unbelievable proportions. There was little doubt that my life was getting ready to change. Forever.

Knowing this, I also knew there were miles to go before I would understand the full meaning and implications of what I had seen. [3]

---

[3] See Illustrations #1 and #2, pp.169-170.

<><><>

CHAPTER 5

A Map

The year was 2009.

I remember waking up from the sleep that had greeted the dream like an old friend. It was then that I heard the words, *"You're to build a University."*

Now it is important that the reader know that as soon as I heard the word "University" I was instantly aware that the word *University* in this case was not being defined as an institute of higher learning in the way that the present day educational culture would define such a facility. What I had seen was uniquely different and of a scope beyond anything, to my knowledge, yet imagined or created. I believe the word *University* may have appeared because at the time of the dream it was the closest thing in my own vocabulary that called out of me a sense of this being a high and noble endeavor.

Hearing the words, *"You're to build a University,"* I saw in my mind's eye the group of buildings seen in my dream, situated on strategic points of our planet. I had a sense at the time that these were the equivalent of planetary acupressure points. From each there was what seemed to be a beacon of light or energy shooting through our atmosphere and into the cosmos.

The next days and weeks were filled with a strange sense of *Knowing*. At the same time, I continued to ponder over what it was that I had been given. I also trusted clarity to the full scope of what I'd seen would come in time.

Let's go further.

What I did know was that whatever it is that I saw was *new*. At first I thought it had to do with *a* new direction in which the concept of *education* would go in the future. Certainly this was part of it. But it was more than that! Much more!

To put it bluntly, the *dream* was a "map." It revealed a way to get from *point A*, being the present state of affairs for humankind, to *point B*, a future that would ensure continuous and positive evolutionary development of human beings that in turn would safeguard that of all life on our planet. This *map* came complete with blueprints and detailed charts.

… I know now that what I had seen was a viable way to help humankind evolve and move into a *sustained state of awareness* that would last the span of an entire lifetime! It was/is a way to plant a seed for communicating more profoundly with ourselves, with each other and all life on Earth *and with life beyond our planet*, a seed that in time would flourish and move the human consciousness into higher states of being.

<>

As the immense scope of the *recurring dream* began coming into focus, I became acutely aware that I will not see its full manifestation in my lifetime. That will be left to others to do. My job is to birth the vision and all this might entail. Perhaps you, the reader, are one of those who will help to bring about the changes necessary to allow what I've seen to unfold and take root. It may even be that you yourself carry a piece of this *dream* deep within your psyche that needs only a catalyst as this simple book to trigger your memory of the part you play in constructing this new future.

CHAPTER 6

Two Parts of a Whole

In 2010 I put out an early Manifesto to the public. It was mostly to announce to a listening Universe that I was saying *yes*. *Yes* that I was willing to move the *dream* I had been given through the birthing canal. *Yes* to doing what I could to bring it into physical reality. I figured that if I had the courage to say it outloud this would provide impetus for my moving forward. It worked.

In the Manifesto I brazenly said the following:

> "The University in my *dream* will incorporate all the principles that call out the best in us, and will be for all age groups, from the youngest to the oldest among us. Academics will look very different and will simply be a byproduct of the true focus, being *the whole person*, encompassing the *mental, emotional, physical* and *spiritual* aspects of who each of us is. The Curriculum will be made up of a variety of experiences, all designed to help expand human potential and awareness, to build deep bonds of emotional and spiritual connectedness that transcend

national identity, ethnic backgrounds and religious beliefs. A primary goal is to create an experience where there is greater understanding, empathy and tolerance of others who are perceived as "different." This will include the learning of new skills for healthier communication and interaction. It is in this way that we will eliminate prejudice, resolve interpersonal and interethnic conflicts, and transcend a world where conflict is too often assumed and indeed encouraged ... All to plant seeds for a world whose underlying premise is one of Love.

"The ultimate goal?? To provide the space and environment that will allow the human species to evolve into an eventual enlightened lifeform."

When I look back and read what I wrote then, I'm reminded of a corporate *mission statement*. I suppose in a way it is very similar if not the same… a statement to declare purpose and direction. The piece, however, I didn't expound upon, or even include, was the *second* part of the *dream* - **communication and healthy interaction with Star Beings**! Visitors from other star systems or dimensions. Quite frankly, both my husband and I thought that sharing this piece might simply turn some people off, especially those of a more pragmatic disposition.

It didn't take me long, however, to see that one part of my dream was intimately intertwined with the other. And that however much my ego wanted to control what I had seen, the fact remained: the *dream* had occurred over and over again and in time came with clear and detailed information. I appreciate that there may be psychologists and psychiatrists who might enjoy dissecting the whole experience. I would simply challenge their choice to stay inside a traditional comfort zone rather than choosing to take a leap and move outside the academic box to see what is possible.

And so, I set out attempting to communicate what I had seen with the hope I wouldn't be viewed as a little nuts. In time, that too, didn't matter.

There was one nagging piece. No matter how I might have described the *dream* in the Manifesto and subsequent documents, intuitively I knew there was *more*. More to come. More of the *dream* still unfolding and revealing itself. My constant companion, with whom I admit I was reluctant to enter into a close relationship, was *patience*.

CHAPTER 7

Five Circles

In January of 2010 I was in Wasserkoog, Germany, a small village situated on the edge of the frigid North Sea where winter winds sing harsh and blustery songs. Here I had an experience that was the catalyst for opening a door, bringing a rush of answers to my many questions that had come over the last weeks since the dream's return.

I was walking down a country road, enjoying the cold breeze, observing the cows and sheep in their close huddles on the flat frozen fields and breathing in the salty scent of the sea nearby. It was a sunny day and my thoughts couldn't seem to go anywhere but to the *recurring dream*.

Finally, at one point, I yelled out to the Universe that I hoped was listening, "Help me understand what I'm seeing. If I'm to help manifest this University (or whatever it is to be called) I NEED MORE INFORMATION!" Clearly, my *patience* was not in great abundance that day. I wasn't shy in vocalizing my frustration of knowing I'd been given something extraordinary *and* that I wanted and needed more detail. The sheep and the cows were listening. I was certainly loud enough!

Suddenly, in my mind's eye, I saw *five large circles*. Each was sitting in a particular direction, as do the five circles of certain mandalas of traditions like those of the Hindu, Buddhist and Native American. Represented were the *directions* of the *East, South, West, North* and the *Center*.

Here was the astounding thing, a true epiphany! The key to unlocking the treasure chest that held the answers to my questions had been in front of me all along. I just hadn't put the pieces together.

For years I had been working with the directional mandalas. What one quickly discovers is that each direction evokes a specific response and experience.[4] For example, facing and focusing on the direction of the East stimulates *mental* capacity and development. Focusing on the South stimulates *emotional* understanding and maturity.

In a like manner, development and appreciation of the *physical* happens in the West. And the unfolding, learning and honing of the *spiritual* occurs in the North.

Then there is the *Center*. The Native American tribe of the Cherokee call the Center, *"where you are"* Those of the

---

[4] The meaning and representation of Directions to Indigenous peoples worldwide varies. The common thread is the idea that all Directions lead to the Great Spirit or to God or to Nirvana, etc., depending on the belief system one adheres to. I, myself, though Cherokee in heritage was introduced to an intimate understanding of the Directional Energies by Native American mystic, Joseph Rael, of the Ute and Picuris tribes in the Southwest of the USA. Subsequently, after years of travel and research I've learned through experience that when focusing in a particular Direction specific responses result. This becomes apparent when singing an individual vowel sound that has been matched to a certain Direction. See Chapter 14 on *Sound*.

Hindu and Buddhist traditions say the Center represents a microcosm of the Universe, the Center being the ultimate state of being. It represents the *Source*. It is a place of *transformation* where tapping into the core of one's self can be experienced. Some, particularly in Chinese culture, have identified this as the place from which the Qi (pronounced *chi* or *chee)* or *life force* moves through all living things. The Hindus call this life force, *prana*. It is called by other names in different cultures.

In Tom Kenyon's book, *The Hathor Material,* the Hathors speak of the *pranic tube,* also called a *prana tube,* an energetic tube that runs from the crown of one's head to the perineum in the lower part of the torso. The *prana* or *life force* moves through this tube and is "the central column of the magnetic field that is emitted by one's physical body." Why all this is important to the vision I'd been given will be made clear when we get into the *Transmissions* in Chapter 17.

Walking down that country road in Germany, I understood. The *five circles* were the five large buildings of the structural complex I had seen in my *dream*.

CHAPTER 8

Willis Talks

As the enormity of the vision seen in my *dream* began to put down roots in my own consciousness, I could feel a fire building inside of me. There was a growing passion to share what I was seeing with anyone who would listen. The Manifesto was one of my first attempts. After that? Well …

What I quickly learned is that though the energy of the zeal to share was enticing enough to light the fires of excitement and curiosity in almost all who would hear me, I found myself stumbling over what seemed like thousands of words trying to pour out of my mouth at the same time, each competing with the other to come out first.

When I made the effort to slow the torrent of words to a pace where a succinct logical order was possible, even then my thoughts felt disconnected and I'm certain were heard by others as a garbled communication of ideas.

I remembered Einstein's quote, "If you can't explain it simply, you don't understand it well." I began to realize how true this statement was. So I started going into meditation asking for help. Shortly after, something rather wonderful happened.

In the years past, one of my dear mentors and friends, Willis Harman, and I would take long walks where we would exchange ideas and engage in rich dialogue on the state of affairs on planet Earth. Most often, however, our conversations would center around the workings, the hows, whys and whats of the universe in which we live and its relationship to other universes and dimensions. These were conversations that fed my soul.

For those not familiar with Willis, he was a well-respected scientist and engineer who easily thought "outside the box" and was known for pushing the envelope of the human experience. He wrote such books as *Global Mind Change* and *An Incomplete Guide to the Future*. He was co-founder of the World Business Academy and President of the Institute of Noetic Sciences, founded by the astronaut Edgar Mitchell.

Willis and I met in the late eighties in Texas at a conference entitled "Consciousness as a Causal Reality." [5]

I grew to cherish the conversations we shared over the years and found them to be catalysts for stimulating my own thinking processes. Thinking *outside the box* was easy for me. Having my ideas explored and confirmed by a person of

---

[5] Details of my first encounter with Willis Harman can be found in our biography *Journey for the One,* written by noted author, Monty Joynes. It was one of those serendipitous events that marks an unexpected milestone in a person's life.

Willis's stature was exactly what I needed in order to take some of the bold steps I did at the time. I began calling these conversations *Willis Talks*.

Willis died in 1997. With him went the coveted moments of our treasured talks and sharing of ideas. He was one of those rare souls who touched the lives of thousands perhaps millions, inspiring so many of us to explore beyond known boundaries. An exceptional human being.

<>

In the late spring of 2012 I think I may have been at the height of frustration, realizing I was having great difficulty putting my *dream* into words that a layperson could understand. This is when the prayers for "help" went out to the ethers.

It was on a pleasant June morning that I decided to go out and sit in our lawn swing (a two-seater) to take a breather and let the gentle breezes clear my head and heart. As I allowed myself to sink into a quiet reverie, I said softly, "I miss our talks, Willis. I need your help in articulating what I've seen. I wish you were here."

Suddenly I heard, "I'm right here." Not only did I hear the words, I also sensed the old familiar grin that was a Willis trademark.

Was it my imagination?? Who knows?! It didn't matter. The impact of this moment was monumental. A door had opened. I literally felt it in every cell of my body, as though a huge breath of fresh oxygen had just entered allowing me to breathe deeply. The experience was one of intense relief. Somehow I knew beyond any doubt, in that moment and all those to follow, a fountain of wisdom was being made available to me that doubtless had been there all along but for some reason I hadn't been able to access. That just changed.[6]

<>

Throughout the summer and early fall, I would sit in the swing and engage in my *Willis Talks*.

At one point I wrote into my Journal:

> "These days seem to be about getting clarity, which includes clearing out the old, preparing for the new. I'm especially enjoying my '*Willis Talks.*' Every day now I've been going to the swing and there Willis asks me questions and there I give the answers I know. He is teaching me to articulate the vision I was given. And yesterday I had a realization. If, indeed, this 'University' is to replace all traditional education then it also means it will be in support of a

---

[6] Some of you reading this may find my experience regarding Willis testing or even defying your world view. If this is the case, "childlike innocence" may be the key. I simply ask you to give yourselves permission to explore possibilities that might challenge the ego but give flight to your imagination.

community whose whole perspective on 'society' is different from what most of us know.

"As an example, people will no longer seek out a profession for the sole purpose of 'making a living,' setting their dreams aside. Rather, it is the *dreams* themselves that come in with each person at the time of birth that are nurtured and manifestation encouraged.

"It seems to be around the age of three years that a person begins to unconsciously ' go to sleep,' spending the rest of her or his life attempting to find the way back to the state of *awareness* he or she came onto the planet with in the first place.[7] Through the concept of what has been shown to me in the *dream* each person receives the opportunity to 'hone and maintain' a conscious state of awareness, in other words, creating a 'wakeful state' for one's total lifetime. In so doing, this creates what is necessary for the human species to evolve to an eventual enlightened lifeform …

---

[7] I realize that using terms like "awareness" can induce certain responses from different corners of the academic world. I merely am pointing out that as human beings, each of us is born with *unlimited potential* however one chooses to define this. This potential, intuitive and otherwise, for most persons seems to go dormant or disappear altogether once the individual begins to interact with a world whose mass consciousness doesn't encourage open exploration and development in this area as being primary to humankind's journey "up the evolutionary ladder" toward an enlightened and fully awake species.

"In this process a *beacon* moves out into the cosmos as a result of the physical structure of the University itself, sending communication signals to our brothers and sisters of this and other star systems.[8] This announces the 'readiness' of our species to interact in an *awakened* state or manner with others off our planet. In a word, authentic and caring communication at all levels not only becomes a possibility, it is manifested into actual physical reality.[9]

"Utopian as all this may sound, it is time. If we are to survive and more importantly, evolve into a higher state of consciousness, then it is necessary to make choices that will allow something as what I have seen in my *dream* to happen. "

More and more I found myself able to verbalize what I had seen. And as I found the words to describe the *dream* I also found that hidden within the thoughts were clues that took me deeper into understanding the many layers and possibilities of what this vision was.

---

[8] Also see Chapter 11, section on *Beacons into Space*

[9] Consider Arthur Clarke's 2001 Space Odyssey in which a scenario is presented of a trigger mechanism (Remember the black monoliths?) that marks the moment humans reach a particular state in their evolutionary development. A signal is set off and moves through the cosmos alerting the advanced civilization who created the alarm system in the first place, that humans are ready to make the next leap in evolution, and subsequently make open contact.

Let's go further, however, into defining first what I had seen and come to know so far, starting with the physical shape of the University and the *directional orientation of the structure itself*.

What follows in the next pages are the beginning stages of what began to reveal itself to me. As the ensuing weeks and months began to roll by, new information started coming. It was as though the Universe was giving me a "bread crumb trail" to follow. I suppose that had I received all the information at once, my brain would have shorted out. Who knows?!

And so, if you will allow me, I'll share the information I received in the same order in which I myself received it, in hopes that you the reader may grasp the depth of what I saw and appreciate the potential that lies inside the vision.

CHAPTER 9

Significance of Thirteen

After my experience on the country road in Germany, seeing the *five circle*s and understanding that they represent the five large circular buildings in the complex seen in my *dream*, I was then able to see and identify the other circular structures of the complex. I was particularly interested in the fact that they totaled a number of *thirteen*. Here is why …

I have learned through our friends in Israel that the number "thirteen" in the Hebrew language also carries the energy and vibration of the number "one." As I believe the *dream* I was given has within it clues to our survival here, and that that survival is dependent upon our ability and willingness to get along with each other and other lifeforms in a positive and healthy way, it would make sense that the number "13" shows up in the way it has. Native Americans would say "We are all related." Another way to put this is, "There is only *one* of us here."

There is an abundance of wonderful books out there on the various theories describing the *connectedness of all things*.[10] John Bell (Bell's Theorem), Rupert Sheldrake (Morphogenetic Fields Theory), physicist David Bohm, Swiss psychologist Carl Jung are a very small representation of the many brave souls who have spent their scientific careers exploring how our universe reflects a single consciousness. A good book for the layperson is Gary Zukav's <u>The Dancing Wu Li Masters.</u> I also recommend Edgar Mitchell's book, <u>Way of the Explorer</u>.

John Pehrson[11], in his new system outlined in his book <u>Mystical Numerology: The Creative Power of Sounds and Numbers</u>, speaks of the number *thirteen* as being about *endings and beginnings*. This is based on the Quiche (K'iche) Mayan calendar day, Aqabal, which is about the *dawn and the dusk*.[12] Relating to my *dream,* something old is falling away and something new is being born.

John also says the number *thirteen* is about the *natural world, about community,* being goal oriented in creating and connecting together. It is the number of *life,* of the *builder*. It is the number to keep the *fabric of society strong*, (including a new kind of society that is alluded to in my *dream).* And on

---

[10] A few of these books are listed at the end of *Eyes Open* in the section of Books of Reference and Excellent Reads.
[11] John, along with Sue Mehrtens, wrote an earlier book, *Intuitive Imagery: A Resource at Work* (publisher Butterworth-Heinemann) which describes a process using intuition as a tool to consciously tap into the part of us that is connected with everything, i.e. the *collective consciousness* of all we perceive to exist.
[12] The Quiche (K'iche) Mayan calendar is made up of twenty days that cycle through *thirteen* powers, (described in detail in John's book on Mystical Numerology).

the mystical level, *thirteen* is about *transcendence.* To support his findings, John has also created a new *13-month Calendar.* There is more but you get the idea.

With the help of the *University,* as more and more of us become conscious of our connection to everyone and everything, as we start learning how to get along with ourselves and honoring other lifeforms on our planet, if my *dream* is correct, our extraterrestrial friends will be willing to come and interact more openly and engage in a level of communication not possible until now. Indeed, I have a feeling they've been waiting a long time for this moment to arrive.

CHAPTER 10

Shape and Orientation

As stated in the last Chapter, the shape of the building complex seen in my *dream* is an overall structure made up of *thirteen circles*. Why *circles*? I believe it is because this shape allows for an unrestricted flow of energy and enhances the vibration or sound of the entire *facility*. More will be said about this in Chapter 14 on *Sound*.

The *thirteen circles* include *five primary buildings, four connecting Bridges, and four Garden areas* interspersed among the five larger buildings. (See Illustration #3, p.171.)

The *shape* of the complex, i.e. *circles*, and the *direction* in which each is placed inside the facility's design are important. Important for two reasons …

First, both the circular shapes and their orientation in specific directions directly support the *kind of learning* that takes place in each *arm* of the University. For example, the direction of the *East* actually does stimulate *mental* capacity and capability; the *South* enhances the emotional self, and so on. This has been tried and proven for centuries with Native

American and other indigenous peoples. And from personal experience as well.

The *second* reason why the physical shape and directional orientation are important is that the entire physical facility of the University itself becomes a BEACON for our extraterrestrial brothers and sisters. This will be enhanced through a foundation of certain kinds of *crystals* embedded into each of the circular structures.

Here is how it will work:

FIVE PRIMARY BUILDINGS AND DIRECTIONAL ORIENTATION

- East / Mental Development:

    One arm of the University is set in the East direction, whose main focus is the MENTAL development of a Human Being. This is where the intellect will be nurtured and stimulated. It is where knowledge traditionally imparted in more common educational systems around the world, will be shared and learned. It is here that the concept of reasoning, of rational thought will be explored. In a word, this is where one learns *how to think*.

- South / Emotional Development:

One arm of the University will be set in the direction of the South. Its focus is the EMOTIONAL development of a Human Being. The levels of this area have multiple possibilities, most of which will be laid out and explored as we get closer to realizing the fulfillment of the University. What I know now is that this area will include delving into the world of emotions, experiencing and understanding them. It also will include the imparting of skills to help individuals interact with others in a healthy way, honoring differences and learning to communicate caringly and authentically. It is here that a person is given the opportunity to come to a better understanding of her/himself and again is given skills to help one traverse the emotional waters when needed.

- West / Physical Development:

One arm of the University will be set in the West. Its focus will be dedicated to the PHYSICAL development of a person. This will comprise what most of us understand as physical education, including individual development as well as team athletic pursuits. However, this arm of the University will go further to include certain kinds of yoga and anything that has to do with developing our awareness of and connection to the physical world around us.

- North / Spiritual Development:

    It is in the arm of the University set in the North where concentration on SPRITUAL development will take place. As an example, this will include the study and practice of Meditation in its many forms. There is much to be said about what will occur in this area and I don't have all the answers. Others will, however. I do know that the focus of this part of the University *is not* about the practice or study of "religion." It is beyond religion. Religion as such will most likely be looked at and studied in the East arm, perhaps as a matter of historical significance and influence.

- Center / Community Development:

    The *Center* of a traditional Native Mandala represents the Source or Transformation or Great Spirit.[13] As regards the University, what I see in the Center building are the very young children and very old adults. This is where they will spend most of their time when at the University. It is here where all in the community will gather from time to time, to share with and learn from the *children* and the *elders*. The young children and the elders are "doorways" for all of us, as the young ones have just *arrived* from the Source and the old ones are *returning*. In both, there is

---

[13] See Chapter 7 for further description of the Center.

an innocence and a wisdom that get buried or forgotten when the Self within us is not nurtured and stimulated in a wholesome and good way.

The Center is a place where skills for being in community with each other in a caring and authentic manner will be taught and honed. **The purpose here is very important, for it is from the Center that the *balance* for the entire community, both within and without the University, will be maintained and nurtured. It is from this place that the seed of how we treat each other will be rooted into the mass consciousness of our planet.**

FOUR BRIDGES AND FOUR GARDEN AREAS

- Bridges:

Each of the four primary arms of the University (East, South, West and North) will be connected to the Center with circular buildings I'm calling *Bridges*. (See Illustration #3, p.171.) They contain certain crystals that will help one adjust to the energy of moving from one arm of the University to another. They may also have small nooks in which to sit and have a quiet moment, study or meditate.

- **Gardens:**

Between each of the four primary buildings or arms of the University are *circular Gardens*, four total. A single Garden will touch two of the outer primaries plus the Center. (See illustration #3, p.171) As there will be four Gardens, one may be dedicated to agricultural pursuits, another to movement and sound as is found in Native ceremonies, another to environmental preservation, and yet another might be a meditation Garden. So many possibilities, each and all dedicated to furthering the purpose of the University.

Those who have had the same *dream* as I have had, as well as other interested and creative persons, will have more detailed input on the *Gardens*. There is also more on the role of the *Gardens* described in Chapters 14 on *Sound*, Chapter 15 on *Crystals* and in various sections of Chapter 17 on *Transmissions*.

Now we go to the *second* part of my *recurring dream* … the part that reads more like *science fiction*. Let us remember though, that it is through the imagination of science fiction that some of the most amazing and groundbreaking inventions and understanding of our present day world have come… all laying the foundation for our possible futures.

<><><>

CHAPTER 11

The ICC (Interstellar Communication Center)

In my *recurring dream,* standing in a hallway where friendly conversation seemed to be taking place between human beings and visitors from off our planet Earth, I was acutely aware that I was looking into a future that is already opening its doors to us.

I realized I was seeing a system and a means to encourage and establish ongoing open communication with beings from our own and other star systems and dimensions. For now, I am calling this particular part of the vision, the *Interstellar Communication Center* or *ICC.*

The physical structure to support this possibility is incorporated into the design of the *University* itself. In other words, both concepts of human development and cultivation of extraterrestrial communication will be found within the same building complex. And as we move through *Eyes Open,* when speaking of the **entire** complex seen in my *recurring dream*, I will most often refer to the *University* and *Interstellar Communication Center* as the ***U/ICC.***

As to involvement of star being visitors, in its initial stages the ICC itself will serve two primary purposes: 1) To establish a base of communication between visitors and humans; and 2) to offer an environment where each might learn from the other and about each other. Both are crucial for a healthy future as an intergalactic community.

Naïve?

Is it naïve to think in such terms? Maybe. I offered this idea to a scientifically minded friend not long ago, who viewed any possible connection with star beings in the way I have seen as being no less than "naïve." I suppose it depends upon from whose perspective one is looking. Perhaps it does require a certain measure of imagination and, as seems to be a prerequisite for new possibilities, the willingness to jump from the edge of that place that is familiar and safe. Consider this. *Boundaries to the imagination are meant to be surpassed…*

However, to step beyond the bounds of the imagination for some, particularly those who require "hard proof" of the mere existence of life beyond Earth, this whole subject may appear to be folly. That even thinking such extraordinary possibility might exist for these individuals may appear to be a flaw in the *creation* process. I would challenge that it is the *ego* afraid to take a step into the unknown…

To those who are willing to stretch out beyond the familiar, delving into the mysterious realm of new possibilities, the existence of *life* beyond our Earth as well as humanity's intimate engaging and interacting with this *life*, may represent the very heartbeat of the act of *creation* itself… all resulting in the birth of something new, wondrous and incredibly beautiful.

There is yet one other aspect around the subject of "naivety" I wish to address here. It has to do with a mindset so prevalent today. One that gets played out in science fiction novels, on television and on the movie screen again and again. And to conspiracy theorists, it is a tool that those in power use to control the collective. That is the *scenario* that we humans *must fear* interaction with "alien beings." That such contact would present a "clear and present danger" and would be a threat to our world and our way of living...

The mindset cultivated then is one that communication with extraterrestrials will undoubtedly produce takeover, manipulation and/or ultimate annihilation of the human race. To think otherwise is considered "naïve."

All I can say is that to believe that if this is the obvious and only outcome resulting from contact with star visitors, then *this* is what is "naïve." And to focus on *fear* rather than the possibility of an enlightened interaction that serves all in a good way is also "naïve."

I'm trusting there are enough souls out there willing, perhaps even driven, to come from a better place within themselves…people who are already creating a bridge upon which the masses may walk into a bigger and more mature way of thinking and being. With this hope and promise, let's continue on…

Language

I'm guessing that most people assume that visitors coming from other systems will be so far advanced that they won't need what might be offered in the ICC.

Technologically this is probably true, the most obvious clue being *that they will have managed to get here in the first place!* They will have made the journey to us from wherever their own homeworld is located, whether arriving via a "flying saucer" traversing light years through a wormhole or space-time portal, or materializing out of one dimension into our own 3-dimensional space.

Where the ICC will come into play is that it will be a place designed to prepare both humans and star visitors to meet and learn about each other. More to the point, it is a place where we might learn how to communicate with each other so that the other is able to *hear and understand.*

Of course, the obvious question is, "How do either of us learn the *language* of the other?" An interesting question as some species may not even have vocal cords or anything similar with which to make sound coherent to the human ear. It, too, is possible that the sounds a human might make in an attempt to communicate would be indecipherable to the "ear" of a star being. Here is where the term *language* will take on a broader perspective…

Sound Communication

All lifeforms emit *sound* in one way or another *through their very existence*,[14] whether or not the *sound* is audible to the human ear. In other words, just because a person cannot hear the *sound* being produced does not mean that the *sound* is not occurring. *The sound vibrations created by any lifeform carry within them the intent and meaning of what is being conveyed.* This is beyond *words*. It is, however, a form of communication, one that precedes or even supersedes formation of words and goes directly to the heart of an idea's essence. For instance, the production of Spontaneous Sound, which requires no thought but only the willingness to allow sound to flow through and from a person, taps into a deep level of consciousness that is the direct line or avenue into the manifestation of a *form* born from an idea.[15] As subtle as this form of communication can be, it is a viable avenue for connecting to another, and though basic in human nature, this way of transmitting and doing so coherently has been lost and is presently undeveloped in most humans, specifically adult humans.

Several instances that come to mind to share regarding communicating intent with the use of *sound* without the use

---

[14] See Chapter 14 on *Sound*. The fact that all is comprised of vibration, the very existence of anything or anyone emits a frequency, a *sound that is the form itself.*
[15] This is even true for persons who do not have functioning vocal cords. The mere intent of sound (silent sound), can produce the same results as that produced by functioning vocal cords. When I had cancer and cured myself with the use of sound and imaging, I produced both audible and silent sound in changing the vibration of the cancer and of my body and found one method to be as powerful as the other.

of words, however elementary, are: 1) the assorted cries of an infant that a parent and/or sibling can decipher in an instant. To the adult this may be experienced as instinctual. To the infant and to the younger children, it is an effective mode of communicating a need or desire. 2) There is my own personal experience of singing with birds. In particular there is a specific mockingbird who during the time of my writing this book *daily* participated in a duet with me, an experience that inevitably drew in an amazing chorus of other feathered friends. The sense of connection was profound. 3) There are multiple well-documented instances of dogs, whales, dolphins, gorillas, apes and innumerable other species, including certain members of the plant kingdom here on Earth who don't speak our *language* but who nonetheless can clearly communicate with humans on a variety of levels. And though some species come to recognize words, I surmise that the actual communication is occurring at a deeper level. 4) And then there's the occurrence of the tonal phrase, familiar to so many of us, that sounded in the movie "Close Encounters of the Third Kind" where humans and star visitors communicated through *sound*, as well as through a *telepathic or psychic* message given to a number of people announcing a designated time and place of the visitors' arrival… a perfect segue to the following section…

Psychic and Intuitive Communication

Perhaps the most obvious example of a broader view on *language* and the one where much, perhaps even most, of our communication with star visitors will happen is *psychically,* or more to the point, *telepathically,* (*telepathy* being one form of *psychic* ability.[16]) The positive of this is that there will be less chance of *mis*understandings to occur. And before I go any further, it might be wise to insert a comment or two here addressing a commonly accepted and unfortunate view, particularly in Western cultures, of both human and alien beings misusing psychic abilities to the point of abuse. Of course, this possibility is always present. But again the premise is that in a highly evolved species, human and otherwise, any beings living and thriving at a lower vibratory state of existence will have a challenge engaging in productive interaction with those beings whose own vibration has raised and increased resulting in a higher state of consciousness. This is explored in more detail in Chapter 19.

As to the practice and perfecting of psychic skills in humans, this will occur in varying degrees for all ages in all arms of the University. And this will happen in tandem with the natural growth and mature development that will take place as a result of all the other exercises and activities made available to everyone in the U/ICC.

---

[16] The transmission of information without the use of any known sensory channel or physical interaction is called *telepathy*. There are several terms used to describe this phenomenon and a variety of sophisticated definitions. What I'm giving you here is a straightforward and simple description. Note that the terms *telepathic* and *psychic* are often used interchangeably, though *telepathy* by definition is a form of *psychic* ability. In simple terms, *psychic* refers to the extrasensory ability to perceive information hidden from the normal senses, to *at will* tap into a person, situation, event, etc. with *intention*, and if desired, communicate at that level.

Also included here is the acknowledging and honing of *intuition* in a person. Intuition is being defined here as "... a form of spontaneous *knowing* without the conscious use of logic or analytical reasoning." [17] Paramahansa Yogananda, beloved by so many for his wisdom as an Indian yogi and guru, defined *Intuition* as *soul guidance,* appearing naturally in a person during those instants when the mind is calm.

I'm reminded of the character Deanna Troi in *Star Trek: The Next Generation* who is half-Betazoid, a species with the unique ability to sense emotions. A term used here when describing this phenomenon and one that is less familiar to most of us, is *psionic.*[18]

Such skills as these are critical to authentic and accurate communication with our own species, and may be more so with off-planet species of life with whom it may appear we have, at least in the onset, very little in common. And the truth is that any of these skills are underdeveloped in most humans yet are accessible by all if an environment can be created to nurture these natural abilities. This is a role for the U/ICC.

Numbers as a Language

Of all the methods of communication, especially in the

---

[17] Webster's Third International Dictionary.
[18] The term *psionic,* most often used in *science fiction,* was coined by John W. Campbell in 1952, from *psi (psyche)* and the ending *onics* from *electronics,* which implied that the paranormal powers of the mind could be made to work reliably. Over the years, its meaning has come to include a combination of both a psychic and intuitive nature.

event of initial encounters with other life beyond Earth, there is one mode not yet mentioned. And I would be remiss not to include such an obvious form of communication. It is the *language of numbers*.

If you will allow me, I want to reference, once again, the wonderful world of the big screen, specifically movies as *Contact* and, as has already been mentioned, *Close Encounters of the Third Kind*.

In *Close Encounters* I spoke earlier of the use of *sound* as an important means of communication. What I didn't say was that *sound* itself can be deciphered into numerical sequences. Why is this true? Because all sounds are made up of specific frequencies and frequencies are described by *numbers*.

Have you ever heard someone say that "music is numbers and vice versa?" Well, if not, you're hearing it now. *Sound,* all *sound*, is comprised of numerical sequences…*numbers!* A rough analogy one might use is, "*Numbers* are to *sound* as *atoms* are to *matter.*" Or "… as *letters* are to *words.*"

In two scenes of Spielberg's movie the use of *numbers* are key in the communication of the alien visitors with humans. The first instance was in the message sent designating the time and place of first contact. The second was the tonal phrasing referred to in the section on the *language of sound* where the familiar 5-notes graduated quickly to an intricate and complicated exchange between the extraterrestrials and humans. This exchange (happening between the mother ship and a humble fella on an elaborate synthesizer) ultimately went beyond a human being's ability to "keep up," requiring the switchover to the more sophisticated capability of a computer. The magic of this scene is that suddenly it was clear that a new language was being introduced, an extraterrestrial language of *numbers through sound*.

In the movie *Contact* we are introduced to Dr. Ellie Arroway, a scientist who has spent her life and career searching for extraterrestrial life. Working for SETI (Search for Extraterrestrial Intelligence)[19] Ellie and her crew spend much of their time sending and monitoring radio signals they actively transmit into space. One of the most poignant scenes happens when radio transmissions are *received* repeating a signal that is immediately recognized as a sequence of *prime numbers*.[20] From there the movie takes off as the intricacies of the *numbers* and their meaning begin to layout blueprints for what is meant to become humanity's first step into becoming a "spacefaring" species.

So how exactly *numbers* as a language will be introduced and incorporated into the workings of the U/ICC is still to be determined. But then… that is part of the adventure

---

[19] The *SETI (Search for Extraterrestrial Intelligence)* projects are a collection of activities currently in place to search out extraterrestrial life. Since 1995 SETI has been funded by private donations. Some of the better known activities are occurring at Harvard University, University of California at Berkeley and the SETI Institute.
[20] A *prime number* is a natural number greater than 1 that can only be divided by 1 and itself. Examples: 1,3,5,7,11,13,17,19,23,29,31 and so on.

we're on, one whose doors are beginning to open wider and wider now.

Size, Shapes, Physical, Non-Physical

It hasn't escaped me that visitors not of Earth will be of varying sizes, shapes, and substances. That indeed, as so many science fiction writers have proposed, many of our star visitors do not exist as *physical* beings but as *nonphysical beings*. Courtney Brown in his fascinating book, Cosmic Voyage, describes the latter as "*subspace beings.*"

I don't pretend to know how all of the details and intricacies necessary to accommodate the many differences and situations that will arise. I do know that what (and who) is needed in the moment will present itself. Our responsibility will be to stay attentive and willing to adapt to the immediate moment and circumstance. In fact, it will be incumbent upon us as the host species to see that this is an attitude ingrained into the spirit of the total U/ICC vision… one of not getting *stuck in form* but one of being *spontaneous*.

And key to this will be how we treat each other as we create this extraordinary experience. Afterall, the *purpose* of the U/ICC and the communities that will surround and support it is to create a reality where all life, regardless of its origin, in its effort to authentically and effectively communicate, does so with integrity and care and as much as is possible, with kindness.

Beacons into Space

I do want to say something here about *beacons*. *Signals* sent out to attract off-planet visitors. Though I will speak of this later in the book,[21] the very structure of the U/ICC – its design and the materials of which it will be made – will emit specific frequencies from Earth into space that will act as beacons, signaling readiness on the part of humans to make contact with other life beyond Earth.

Undoubtedly, there are an unimaginable number of "intelligent" species in the cosmos, many of whom will be able to receive and identify a signal or beacon that we might send out. The crux of the matter is this. *What will be the interpretation of any signal we might transmit?* To quote Edgar Mitchell from *The Way of the Explorer*, "Managing information intentionally is, of course, a wonderful definition of intelligence. There, we cannot state that the meaning of information is contained in the signal. Even if it were, there is no assurance that any receiving entity will perceive its meaning."

Therefore, if I accept Mitchell's premise, it makes sense to me that those who are most likely to be *early* responders to a beacon we might emit are those who have purposely and by design been waiting for humankind to reach a certain stage of evolution and maturity.

This prompts an obvious question. *Why*??

---

[21] Chapter 13: *Pyramids within Pyramids*.

And the inevitable question that then follows, asked by innumerable people who believe visitors have been coming to Earth for eons, "Why are they coming *here*?"

I'm guessing that for as many people who have pondered this question there are as many answers. However a person arrives at her or his answer, the conclusion usually goes one of four ways…

Either such interstellar visits are speculated to be malevolent aimed with aggressive intentions of somehow overtaking, enslaving, or demolishing humans as stated earlier in this chapter. Again all a favorite theme of science fiction writers and film producers. This includes the scenario that our very existence may be the result of our having been created by a "master race of alien beings" for such sinister reasons as explored in the movies *Stargate* and *Matrix*.

Or their visits are as "cosmic curiosity seekers" whose intentions are ill-defined.

Neither of these answers quite frankly is romantic or noble. Nor do they inspire us to reach for the highest of who we are and what we know.

… At least this seems to be true unless the *curiosity seekers* are actually *space explorers* seeking out new worlds as has been done for decades in Gene Roddenberry's *Star Trek*. Entering through the door of science fiction, Roddenberry has continued to spark the imagination of millions, inspiring humanity to "reach for the stars" and consider the possibility of space exploration as an honorable quest. I can even remember as a young child being asked the age old question, "What do you want to be when you grow up?" Always my answer was, "An explorer," and shortly after I would add, "…in outer space."[22]

A *third* choice of why interstellar visitors might be coming here is that they are cosmic beings benevolent in nature, who monitor and assist life on Earth when it seems as though development and growth can benefit from such help. This particular scenario has been carried down through the ages through legend and literature, particularly in Greek, Roman, and Egyptian mythology.

Then there is a *fourth* choice and the one that for me rings truer than all the other possibilities put together, those stated and unstated...

Though all the answers given thusfar have validity, some more than others, for me the answer is fairly simple. I have a sense there is something in our own DNA and/or that of other lifeforms on Earth that is directly related to those who come from the stars; that those of us on Earth may represent some element up or down the evolutionary ladder that has some bearing on the past or future of certain interstellar species of life. Barbara Marciniak in her book *Bringers of the*

---

[22] I'm sure my parents would remember how often I would make the bold statement, "I want to be the first woman in outer space." Though my path took a different turn in the immediate years that followed, I remember how thrilled and awed I was that on June 16, 1963 Russian cosmonaut, *Valentina Tereshkova*, became the first woman to travel into space.

Dawn offers an interesting, if imaginative possibility suggesting Earth itself is a kind of *library* into which species throughout the cosmos have invested some example of their homeworld from which all others throughout the universes may observe and learn. In this way, Earth is also like an Ark. It is precious and is to be watched over as it may very well carry the seeds for all life everywhere.

Now what I have just shared may seem a bit "out there" for some of you. Maybe not. Again for whatever reason our interstellar brothers and sisters want to come here, the occurrence is *inevitable* for our near future. This is a certainty, whether or not there have been visits in the past, and whether or not we may already have star beings who have been living with us for centuries.

CHAPTER 12

"Universe Cities"

While in Germany in 2010, I shared the vision of my *dream* with an extraordinary group of people. There were some twenty-five individuals from different countries around the world who had come together to explore and experience in a microcosmic setting what the world might be like if we humans were committed to communicating in a genuine, honest and caring way. The enthusiastic acceptance from the group of what I showed to them was confirmation telling me that the time was nearing for turning my own focus and energy into bringing my *recurring dream* into physical reality.

At dinner that evening, situated in a beautiful dining room with one long table that allowed seating for the whole group, I heard loud raucous laughter coming from one end of the table. I looked up to see five or six women laughing, seemingly proud of themselves for having an epiphany they were excited to share with me. It had to do with the name of the multiplex facility seen in my *dream*.

I had been calling it the " University," knowing full well this might not be the name that ends up identifying what I had seen. Though the directive "You're to build a

University" came in 2009, even then and as I have already stated, I was aware it was a name to "meet" me where I was at the time in my understanding of the *dream's* size and concept.

And so the small lot of ladies, pleased with their idea, came to me and said, "We think you ought to call the University, *Universe City.*" We all laughed. However, as the next days, weeks and months rolled on, I found the name stuck. But it was not to replace the name of the University. It was for something else. Something that didn't reveal itself until the *"Willis Talks"* began.

It was during the summer of 2012, sitting in the swing, *conversing* with Willis, that I became aware of a piece of the *dream* I hadn't seen before. Surrounding the U/ICC in a circular fashion was a *community*. I remember how startled I felt, wondering why I hadn't seen this in the first place. It wasn't something I initially remember having seen in my *recurring dream* but I always had a sense of its being there. It was almost like a veil had been lightly covering the *dream's* edges that only now was being lifted.

Once I saw this, the information started flooding in. ...

Initially such a community will be comprised of between five hundred and one thousand people. It will serve as a prototype for communities worldwide as present-day systems begin to break down and necessity forces humankind to explore, *on a massive scale*, alternative ways of living.

I am affectionately calling these communities, *"Universe Cities,"* thanks to our creative dinner group in Germany.

The body of the U/ICC will consist of the members of the encircling community. In fact, each element is crucial for the other to exist as it is a symbiotic relationship.

In the beginning there will be three to five U/ICC sites built on strategic points on Earth, similar to planetary acupressure points. To start with, each will exist on a different continent. This will include surrounding "Universe Cities" made up of five-hundred to a thousand persons *per site*. Volunteers all. Volunteers committed to supporting and bringing what I have seen in the *dream* into existence.

Granted, there are many details to be worked out to make all this possible, but that will come as those who are part of the *dream* step forward. It's important to remember **there are at least twelve more of you out there** who have had some part of this *same dream*, each with clues as to how all of this will be brought into physical being.

Lest you forget, I'm writing this book to find you.

CHAPTER 13

Pyramids within Pyramids

Late one night not long after I had begun writing *Eyes Open,* I was awakened suddenly from a deep sleep. Now, I like to sleep in a space that is pitch-black. I think because it reminds me of the vastness of space and all the potential and wonder that promises. I tell you this because of what I saw. It happened in the blink of an eye. It was crystal clear in both shape and color. And it was the depth of the darkness that allowed the geometric form to stand out in sharp contrast.

What I saw was being observed from an aerial view which at first made the figure look flat. This was until I realized that if seen from the side, it would appear three-dimensional or holographic. It was a 4-sided square *within* another 4-sided square located *within* yet another 4-sided square.

All were somehow coming to a common point in the center. Each of the sides seemed to be lined with smaller squares, producing a honeycomb-like effect. The whole design glowed a bright gold.

The next day I drew what I had seen. I placed it on a wall where I could see the shape any time I walked by. It was placed near a drawing I had made of the thirteen circles described in Chapter 9. (See Illustrations #4 and #5, p.172 .)

Something kept pulling at me every time I walked by the two drawings, but I couldn't identify what it was.

Then one evening sitting in my easy chair in the living room, still able to see the drawings in the other room, I turned on the television to find there was a program on *pyramids*. To borrow a phrase from Bob Hoskins in the movie *Hook,* "Lightning struck my brain!" If I'd been in a cartoon drawing, the cartoonist might have drawn a light bulb flashing over my head in the moment I put two and two together. I had the unexpected realization of the image given me a few nights before as being one of multiple pyramids! This was incredible enough. But when I suddenly found myself superimposing the image of the 4-sided pyramidal squares onto the circular design of the U/ICC, I was taken aback! The pyramid design fit perfectly into the overall *thirteen circles*.

I realize, too, that what I was seeing weren't *physical solid pyramids* but were *energy pyramids!*

It looked like this: The *first energy pyramid* was on the outside connecting the four primary outer buildings to a central point in the Center building. Then there was a *second pyramid* connecting the four Garden areas, again, to the central point of the Center building. And finally there was a *third pyramid* connecting the four Bridges also to the central point in the Center building. (See illustration #6, p.172.)

This information in itself is fascinating but where it is particularly significant is in the fact that crystals are an important part of the U/ICC design. If certain crystals are put into the pyramidal shapes already designated by the structure's circular pattern, then what we have is the creation of *energy pyramids* producing *cosmic beacon lights or signals.*[23] (See Illustration #7, p.173.)

One of the sketches I drew out months ago was one of the Earth with beams of energy being released from specific points on the planet. One of the theories of the ancient pyramid shapes and their precise placement on Earth was that they sent signals into the universe as a way of communicating with those on other planets and stars.

When I grasped this one of many pieces of the puzzle in my *recurring dream*, I was elated! In Chapter 11 I spoke of the second part of my *dream* and that the shape of the building design itself would make it possible for "*beacons*" to go into space, acting as a means to contact our star relatives. I didn't understand at the time that *energy pyramids* were central to the design of the U/ICC and were already included in its *blueprint.*

---

[23] This is also discussed in more detail in Chapters 15 and 17.

Again, this information had been there all along. I just hadn't seen it until now. And whoever it was that sent the image I had that night when I awoke to see the gold "squares," a hearty thank you!

CHAPTER 14

Sound

The next topic is one that will be on intimate terms with all aspects of the U/ICC. Yes, it along with all its many faces and factors will be explored as part of the *curriculum*. However, I'm speaking here in a much broader sense. The topic is *Sound* and the role it will play.

In Chapter 11 we looked at the way in which *sound* is used in the act of communication. I want to expound on this and present an even wider perspective, particularly as it relates to the purpose and task of the U/ICC.

Explore with me for a moment, and if need be, give yourself permission to see through different eyes… or in this case, *listen* with different ears…

Consider that everything we perceive to exist is *vibration*. Another way to say this is that everything is *sound*. And if you will allow me, in this instance I am going to use the word *sound* and *vibration* interchangeably. Consider that all things, be they ideas or feelings or physical matter or of a non-physical state, be they molecules, atoms…that absolutely everything we imagine or believe to exist is comprised of a specific vibratory or *sound signature*. Such signatures are what give identity to everything we have perceived in our reality. Even the very *idea* of *reality* itself is a *sound* that creates what we identify as "reality." The same can be said of the concept of a tree, or a cup of coffee, or a feeling of joy, or a thought or emotion of any sort. Each is a *vibration,* unique unto itself and to which we humans have assigned a name and an identity.

Allow yourself to see that all *sound* is *music*. If this is so, then everything in effect is a *song*. Elegant, isn't it?

Many years ago my husband and I were in Australia where we did a number of workshops and concerts. It was here that I first heard the phrase *"singing ourselves into existence."* This was an idea I'd never encountered before. It was from this perspective, we learned, that the humble Australian Aborigine lived life. Clearly it was a unique view of life and a shared wisdom that to this day has influenced my own perception of reality. In a short time, I began to find it quite easy to think of myself, everyone and everything around me as "singing ourselves into existence." What a simple way to illustrate an example of the makeup of our universe… that everything is vibration. Everything a song, thus everything through the very idea of its existence is "singing itself."

Think about it. The idea itself is pretty profound. If you are willing to allow yourself to move way "outside the box" and make the attempt to *listen* to all around you, then what you hear is an entire cosmic symphony, the ultimate musical masterpiece.

I remember when I first began giving sound workshops. One of the exercises included was the practice of *listening to everything* around you, from the most spectacular to the very mundane. Understand, this is about *listening* with more than physical ears. It is about being, at will, completely and utterly *aware* and *awake*. This may take some practice but it not only is possible, it is necessary if one is to interact with truth and legitimacy inside the depth of one's reality. As this is primary to the purpose and existence of the U/ICC, I hope you can see why focus on this subject deserves some attention.

Once, in one of my sound workshops in the hills outside Haifa, Israel I asked a group to *sing the sounds* they were "hearing" when they let themselves become aware of the scene around them. I suggested they use their imagination if they weren't certain of *hearing* the vibration or sound of a thing, which included colors and shapes as well as objects themselves. Being in the home of an exceptional artist whose house was decorated with his numerous paintings, everyone seemed to home in on the vivid and varied aspects inside these wonderful images.

When I asked those in the group to then move their attention toward the window and focus on what they were seeing outside, one couldn't help but notice the overwhelming abundance of rocks … rocks of all sizes, most of which were the size of a fist or smaller. And so, if indeed each of these stones represented the vibration or song of itself, then the entire hillside was sounding an ongoing massive concert, the beauty of which was able to be accessed if a person were willing to *listen*.

When you add the symphony of the stones to that of the paintings and expand that on out to include each item in our host's home, including ourselves … and then, let this expand out to include every idea, every thought, every emotion, everything and everyone inside Israel …and continue on, moving out beyond Israel to include the Earth … expanding this out to include our universe and beyond what we know and think we know, entering into our imagination … *then* we begin to have some appreciation for the ongoing musical masterpiece that is being played out through the vibrations, the *sounds* that compose and create us all.

I realize I've gotten somewhat poetic here but as the book's author I have the license to play a bit, yes? The point is, whether one uses scientific terms, philosophical or poetic terms to describe the experience of existence, everything has a vibrational signature, each being unique unto itself. In the Aborigine's terms, "everything is singing."

So, in terms of the U/ICC, the fact that the very design of the overall structure itself *is* a specific vibratory signature is not only interesting, it is of particular importance. And the fact that each component of the overall structure is within itself a unique *sound signature* is of added importance. Here is why…

As each of the directions (*East, South, West* and *North*) has a distinct meaning, which has already been described in Chapter 10, each too has a specific *sound*. The sound of the *East* is *AH*. The sound of the *South* is *EH*. The sound of the *West* is *EE*. The *North* is *OH*. And finally the sound of the *Center* is *OO*.

If you notice, these are the *five primary vowel sounds, A (ah), E (eh), I (ee), O (oh), and U (oo)*.[24] They are said by multiple indigenous cultures to be the fundamental vibrations from which everything comes. The vowels themselves are called *principle ideas*. They, in all their different myriad of possibilities, are the essence that create the sound identification of all things.

As each building in the U/ICC is standing in a particular direction in relation to the whole complex itself, each resonates a unique and specific underlying vibration.

For example, the primary building situated in the East whose focus is that of the *mental,* through its very positioning resonates the vibration or sound of *AH*. That sound then, subtle as it is, stimulates the mental capacity and capabilities innate to all humans.

---

[24] The five vowels mentioned here are pronounced as they would be in Spanish, Hebrew and a number of other languages. There are some languages, however, that have some combinations of these vowels sounds representing a single "vowel," as in English. For example, if one listens closely you will hear that the sound of the long A in English is actually a combination of two sounds, "eh, ee,"(E,I)). The sound of "U" in English is a combination of "ee, oo,"(I,U), etc. This is why we suggest pronouncing the five primary vowel sounds as they would be spoken in Spanish or Hebrew where each vowel sound remains pure.

Note that each of the *Gardens* will be situated *between* two of the larger primary buildings. (See illustration #3, p.171 .) In other words the *Garden* that stands between the East and the South will create the vibration of both directions, *AH* and *EH*. These vibrations will resonate within the fundamental structure of that particular *Garden*. As the East is about the *mental* and the South is about the *emotional* self, whatever takes place in that *Garden* will include activities that are designed to be in harmony with the resonance of those directions.

Spontaneous Sound

Here is a good place to introduce what I call *Spontaneous Sound or Singing*. I include this subject because it will be an integral part of the curriculum and other aspects of the U/ICC.

The reasons become obvious once a person or group experiences the results of this phenomenon. In a word, singing sounds spontaneously, *beginning with the vowel sounds*, creating sounds with no words and *no thinking*, transcends the ego and expands and awakens the deeper levels of consciousness. *It is from this place that ideas can be manifested into physical reality.* This includes the reshaping of one's own perception of reality, healing from emotional or physical imbalance for both the individual and the collective. It, too, is a powerful effective way to communicate when words and thoughts are a hindrance.[25]

To give some clarity, let me share a few paragraphs from an article I wrote several years ago:

> "In a vision, in 1996, I was shown that when we sing the vowel sounds, the sounds of the directions of the *East, South, West, North,* and *Center* [A (Ah), E (Eh), I (Ee), O (Oh), U (Oo)], by allowing them to flow ***spontaneously***, we create vibrations that change us all, that help us all remember who we really are and where we come from.
>
> "It was in 1998, living in the San Juan Mountains of Colorado, that, in one week, I had three dreams… With the first, I was standing before some 40,000 people in a stadium, teaching them to sing spontaneously, each person singing who she or he was. You would think the sound would be chaos. It was! But, what happens when you leave something in chaos long enough? The law of physics says it will find its way to order…and in this case, to harmony. In

---

[25] My own story of healing from cancer with the use of sound can be found in our biography, *Journey for the One* by Monty Joynes.

the dream, I could *hear* what would happen. The mass of voices did find their way to harmony. It was the sound of the heavens opening up. It was indescribably beautiful!

"The second dream was similar. I was standing on a stage in an auditorium seated with hundreds of people...again, teaching them to sing in this way. The impact was overwhelming in its power and beauty. [This has happened now, many times over the years since my having the dream.]

"Then came the third dream. I was standing in the United Nations. I had asked each of the delegates, representing countries from all over the world to stand and sing who they were. ***Note: when one sings the sounds of the directions, allowing the self to go into "no mind" or to slip "between the slices of light"(to use a favorite phrase of Native mystic Joseph Rael) the ego is not involved.*** Thus, when the UN delegates began to create spontaneously, from an egoless state, I heard a miracle. I heard the sound of

Peace. I heard the sound of an entire planet remembering who it was. It was the vibration of absolute truth…the remembrance that there is only one person here and that all there is, is love.

"If this whole concept sounds impractical or unbelievable, try it yourself. Create your own songs. Above all, **don't think!** Start with the fundamental sounds of "Ah, Eh, Ee, Oh, Oo," in whatever order they naturally want to come. Then, observe.

"Indigenous peoples have done this for centuries. It is in all of us to sing in this way. It's just that most of us have forgotten. It's time now to remember."

Crystals and Sound

One other key piece to mention here is involvement of *crystals* and *sound*. As Crystals will play an important role in a variety of ways inside the structure and function of the U/ICC I've given the next chapter over to this subject.

Note: More information regarding both Sound and crystals also came through The Transmissions, (Chapter 17), as you will soon see.

<>

Perhaps nothing more needs to be said at this point other than to reiterate the strategic and fundamental role of *sound* as the U/ICC becomes a physical reality.[26]

<><><>

---

[26] Three excellent sources regarding *sound* can be found in Joseph Rael's book of the same name, *Sound* , his earlier book, *Being and Vibration,* and John Pehrson's book, *Mystical Numerology: The Creative Power of Sounds and Numbers* mentioned in Chapter 9. Each goes into great detail on the sounds and meanings of the directions.

CHAPTER 15

Crystals

Many believe crystals themselves to be living lifeforms. My first experience that gave me a sense that this statement was true happened one evening when I was singing in what is called a Peace Sound Chamber, a structure built in accordance with a vision seen by mystic Joseph Rael, many years ago.

Such a chamber is similar to the ancient Native kivas, oval shape, part in the ground, part out. On this particular occasion I was sitting in the center of one of these chambers and was surrounded by various crystals, some of which were inside the walls of the chamber, some in the ground and some loose sitting in numerous places. I was alone inside the chamber. I began to sing spontaneous songs as described in Chapter 14. Shortly after I began, I started to notice something that I couldn't identify at first. It started out as a high pitch almost too subtle for the human ear. The more I sang, however, the more I noticed that harmonies began to develop from *outside* of me. At first, I had no idea where the sounds were coming from. I surmised they might even be angels or other (musical) spirits in the chamber with me.

Suddenly, I started noticing patterns develop. When the sounds continued even after I had stopped singing, I had a surprising realization. The *crystals* themselves were singing. Some might say they were simply resonating *sympathetic vibrations* to the sounds I had made. Yet, again they were creating ongoing patterns of *sound* even after I had stopped singing. Subtle. But nonetheless, still there. From that time on and to this day, I'm convinced that I was encountering a viable lifeform, one that is uniquely different from a human being.

Were the *crystals* communicating with me? I would like to think they were. I suppose those grounded in traditional physics might suggest that the *crystals* were simply demonstrating a resonance stream that once started (the vibrations of my singing acting as a catalyst) continued as the sound waves bounced from one *crystal* to another. I'm choosing to believe it was more than this. My intuition said that it was more. That something within the being of the *crystals* was responding to another being not of its kind. In effect, a kind of communication was taking place.

Now whether a relationship between humans and *crystals* can be established may be a subjective experience, not unlike what takes place between a tree and a person, a special place and a person, etc. I simply know that something out of my ordinary perception of reality occurred and this was exciting. It carries the promise that the premise that "all life is connected" is more than a philosophical supposition, as has been mentioned by John Bell, Rupert Sheldrake, David Bohm and so many others who have dedicated their lives to reveal and confirm.[27]

Role of Crystals in the U/ICC

Regardless of one's bent or understanding of the possibility of a *crystal* as a sentient being, we do know that crystals respond to *sound,* reflecting even absorbing it. Not only do they reflect *sound,* they enhance and expand it. These characteristics are crucial for a couple of reasons:

First, crystals will magnify and help to plant and confirm the experiences gleaned in the U/ICC in relation to what is imparted through the *curriculum* and *purpose* of the U/ICC. Which crystals are better for each building and Garden as each relates to a particular direction will be determined in the initial building plans of the total structure, and by those much wiser than me.

A *second* role that crystals play will be to work with the directions (*East, South, West, North,* and *Center*) to create an identifiable cosmic *beacon* or signal needed for the ICC to communicate beyond our planet. This will be in conjunction with the pyramidal energy described in Chapter 13.

---

[27] Over the years I have continued to explore and cultivate a relationship with crystals. Though I may not have indepth knowledge of the different kinds and nature of many crystals, I have enjoyed discovering a world I might not have even thought possible were it not for my early experience inside the Peace Sound Chamber. Also see Chapter 9 referencing Bell, Sheldrake, and Bohm.

Then there is yet *another* task that crystals will perform. Certain kinds of crystals through their frequency and internal structure and substance, assist to create *space-time portals*, a method of travel for certain species. Not far into the future, this will include human beings as well.

Again, more will be said on *crystals* in Chapter 17, information that came through the *Transmissions*. To use a favored cliché, the best is yet to come.

PART III

The Transmissions

*"When someone stretches out a hand,
  Don't stop to look if it's green.
 Take the hand."*
               - Documentary on Star Wars

CHAPTER 16

Prelude

Recently, before starting this book, I was told by a seer friend of ours that I was going to start receiving *transmissions*. This information was going to come from the stars and would help to give further clarity on the *recurring dream* I had received.

I realize that the very term "transmissions" in the context in which I now use it, can throw up red flags for some individuals. If this is the case, I ask you to give yourself the gift of flying with me. *Go the distance* and read on.

<>

Having set aside a number of months to write *Eyes Open,* I created a living space to support my writing process, affectionately calling it my "writing lab." I also purified my diet and incorporated certain exercises and meditations into my daily schedule to clear and prepare my mind and body. Then I proceeded to write.

It didn't take long to become aware that *new* information was coming in and sometimes so quickly I could hardly get it down fast enough.[28] (Imagination or not, I could sense the comforting presence and support of beings I couldn't always see but was acutely aware of their being within my vibrational field. This was especially true of a little Gray I call "JJ" that I encountered early on and a particular group of individuals that from time to time projected a luminous light maybe for no other reason than to let me know I wasn't journeying into this project by myself. This was helpful as I really did feel as though I was heading into unknown territory.)

And silly as this might sound to some, the stubborn constant flicker of a light or the sudden song of a bird that just landed on the window sill, (inches from my writing table!), became two of several signs that I began to understand were signals to stop whatever I was doing and take notice, even if it meant stopping in the middle of writing a sentence for this book. Other such occurrences would happen when in my sleep someone would appear in a dream telling me to wake up. Or I would be awakened by a sudden sound or light. Once I got it that something or someone was attempting to get my attention, I would fetch my pad and pen, sit in a quiet place and let the information come. And come it did.

---

[28] I have deep appreciation for those like Ken Carey who wrote *Starseed Transmissions,* Tom Kenyon who wrote *The Hathor Material,* and so many others who have stepped out on a limb to share information they believe to have come in from sources outside of earth.

And though what I've included here is a compilation of information received up to the time of this book's printing, I continue to receive data, all of which is being recorded and documented.

I've debated on how best to share in *Eyes Open* what I have received in the past months. At first, I thought to convey the information in the order in which it was received, including dates and times. This would have been an exotic way to do it since most of the clock-times were at some ungodly hour in the night, usually in the wee hours of early morning. But in retrospect it made more sense to present the transmitted material by topic as I recognized that pieces of a single topic would appear on different days.

I have also attempted to put the information received into a language that persons of varied backgrounds and inclinations will, hopefully, be able to hear.

Here goes …

CHAPTER 17

Transmissions

Note: There are 42 Transmissions included here. Because of the brevity of each, all but two, Transit System (Chapter 18) and Love (Chapter 19), have been listed and described in this single Chapter. The 42 are addressed in this order:

1) Recognizing Others Who have had the Dream
2) Multi-Dimensional
3) Five Levels
4) Underground/Underwater Effect
5) Learning for Humans
6) Indoctrination and Adaptation For Off-Planet Species
7) Lower Level and Conditioning areas
8) Center of Top Level
9) Center Building's Role in Energy Flow for the Entire Facility
10) Sound
11) Building Materials
12) Angles
13) Crystals
14) Streams of Water
23) More on Children's Games
24) Consciousness
25) Changing Universe
26) Hundredth Monkey
27) Vibration Increase
28) Conducting Garments
29) Faraday Spaces and Silence
30) Conversation Rooms and Active Listening
31) Lighting
32) Breathable Air
33) Colors and Patterns
34) Places for Nourishment (Cafeterias, etc.)
35) Bathrooms (Accommodations for disposing of waste material

15) Plant Kingdom
16) Animal Kingdom
17) Balance
18) Tunnels
 (described in Chapter 18)
19) Transit System
20) Extrasensory Skills
21) Open Heart and Atlantis
22) More on Children
36) Simultaneous Learning
37) Twenty-Four Hour
 Curriculum
38) Relationship to
 Water, Land, Air
39) Gardening
40) The Five Elements
41) Vibration of Love
 (Chapter 19)
42) Monitoring Council

<>

- **Recognizing others who have had the dream:**

The *recurring dream* is a *dream* for all peoples, though most will not appreciate this in the beginning. The individuals who have had some portion of this *dream* are being guided toward each other. *Immediate recognition will take place. It is important to stay aware and attentive.* Not all those who have had the *dream* are from Earth, a fact that may be hard for a general (and skeptical) public to accept at first.

- **Multi-Dimensional:**

The U/ICC has several dimensions to it, both physical and non-physical. Many already have some appreciation of the different levels of energy that often are identified as existing in *astral* form or in *subspace*. However one describes this phenomenon, the U/ICC will exist on more than the physical plane.

This will become noticeably clear once the facility begins to be built.

It is helpful to be in a *spherical* or *holographic mindset* when imagining the possibility of all this. Not only does this make it simpler to envision the *whole* of this concept rather than mere fragmented pieces, it will also make it easier to grasp the idea of dimensions beyond *physical reality*.

It is important not to limit oneself and the perception of what has the possibility of being. There are exercises, such as the Hathor *Geometries*[29], that can help shift, strengthen and amplify the consciousness of a person and help her/him to be more receptive to see what is possible.

Referencing the overall structure being *multi-dimensional*, the difference in the present day collective consciousness or state of awareness regarding the physical, astral and ethereal levels is that this state of awareness at the time the U/ICC comes into being, will be accepted and understood among the masses. At the time this book is being written there are only a few willing to accept this possibility and who understand the meaning of what this all entails.

---

[29] The *Geometries* involve the movement of a ball of light within the shape of an infinity sign and the moving pattern of an atom. It also includes various exercises involving the figure of an *octahedron,* (An 8-sided polyhedron. Imagine the shape of two square based pyramids joined at the base). See Tom Kenyon's book, *The Hathor Material.*

However, it is important to know that the time for a mass consciousness shift at this level is *imminent*. Not centuries or even decades away. This shift has, in fact, already begun.

- **Five Levels:**

There will be a minimum of *five levels* to the five primary buildings and four Bridges (described in Chapter 10.) Underneath each level of the *five primaries* lies a foundation conducive to supporting the curriculum and enhancing the learning/training that takes place inside that particular level. The specialness of the *foundation* lies in the architectural design and the physical material of which the buildings themselves will be comprised.

The many facets and qualities that make up the human species, involving the *Mental, Emotional, Physical, Spiritual* and the *Center (Source),* are the essential elements around which the physical structure of the U/ICC is to be designed.

Within each level are the key pieces that open the door to *communication development* that will take place among species. In the initial stages of the U/ICC this will first occur between humans themselves; then eventually between humans and interstellar as well as interdimensional beings.

- **Underground/Underwater**:

   There will be multiple U/ICC facilities built around the world, along with accompanying communities or Universe Cities. The geographic characteristics of a particular location will determine whether part of the U/ICC itself will be built *underground* or *under water*. One or the other is necessary to complete the task of the ICC in particular.

- **Learning for Humans**:

   The initial learning for humans will occur throughout the *above ground* part of the facility.[30] Though *eventual* interaction with other species from off-planet will be included here, the immediate learning will involve multi-cultural interaction of humans *with each other* plus human involvement with other of earth's species, both of the plant and animal kingdoms… in particular *water* and *land animals* whose own stages of evolution evoke *love* and emotions of empathy.

   Taking "baby steps" in the beginning stages of humanity's shift and elevation into a higher consciousness will require patience. This is necessary, however, if life on Earth is to evolve in a healthy and

---

[30] This is true except for the very top level of the Center Building, described later in this Chapter.

good way, and if a strong and wise foundation is to be built for a future spacefaring species.

- **<u>Indoctrination and Adaptation for Off-Planet Species:</u>**

  All species coming to Earth will have gone through certain conditioning and/or training before arriving, understanding that many have been coming here for centuries and most likely have already adapted. As will become apparent, however, all coming in the future, traveling from and into *physical space*, conditioned or not, trained or not, will require some adjustment. Here is why:

  > Numerous extraterrestrials that have made contact with humans in the past have done so by providing an experience of frequency alteration affecting the human body which results in a vibrational shift, all with the purpose of assisting humankind in setting the stage for a next step up the evolutionary ladder.

  > As the U/ICC comes to life, however, such star visitors, even though their very presence will continue to have some effect on the frequency of a human being, will themselves, along with other visiting star beings, choose to adjust their own frequency to adapt more

readily to that of Earth, rather than the other way around as has occurred in the past.

The reasons here for all the varieties of fine-tuning and modification are to cultivate an atmosphere for open and trustworthy communication among the many intergalactic species. In the earlier phases of human development this was not possible, but as humankind has evolved over the millennia, the frequency of the species itself has been changing. Indeed the vibration is raising. This has made it possible for open viable contact to take place with other species already of a similar vibratory field and those able and willing to shift to a like vibration. Authentic interaction, including the building of trust, with interested star visitors not only is now achievable, it is advisable as the human race becomes part of a spacefaring community.

- **Lower Level and Conditioning Areas:**

As part of the U/ICC will be underground or underwater, this area will be given predominantly to the welcoming and indoctrination of visitors from other planets and systems. This will include "conditioning areas" to help assist those who may need to adjust to our planet's atmosphere and temperatures. This will not appear as difficult as one might think at first, because in order for a lifeform to exist in a material state, there is an acceptable range of temperatures that make this possible, as well as other conditions and inbuilt mechanisms unique to existing in a physical environment. There is a good example that already exists on Earth demonstrating diversity and adaptability where there are cold-blooded and warm-blooded beings existing at the same time. As long as freezing or burning do not occur and the atmosphere is friendly, different lifeforms can find, within some limits, compatibility.

> - *Why Underground:* One main reason for being underground initially is to allow for those needing to adjust to the *light* provided by the sun to do so safely. (Understand, however, that there are many beings whose own presence emits light that may for humans be near to blinding in intensity. In these instances, it will not matter whether such beings are in an *underground* environment as their adjustment will be in the shifting of their own "bodily" frequency to *tune* to that of Earth.)

> - *Grounding:* An added and important reason why our star visitors will first undergo adjustments in the lower area of the U/ICC has to do with *grounding*. With the help of the Earth itself, those not of this planet can be served by drawing into

themselves the life force energy of the Earth, thus grounding themselves into Earth's vibration and into this reality. By being grounded in Earth's energy, those from other systems can more easily and accurately respond to situations encountered on this planet. There are already a number of tried and true meditations to help with this though others who have had this *dream* will have further detail on how all the above is to transpire.

- *Breathing Oxygen:* For off-world visitors who are unaccustomed to *breathing oxygen (i.e. Earth's air)*, assistance and specific apparatuses will be designed and provided.

- *Adjustments for Subspace Beings:* As to lifeforms traveling *interdimensionally,* adjustments will occur at a *molecular* level in most cases. This is a common mode of travel among these beings. For such *subspace beings* who have the ability to shift between the physical and non-physical, they will also have the ability to naturally adjust themselves to both atmosphere and temperature.

- *Gravity:* Adjustments to the sensation of density caused by Earth's *gravity* will be necessary and specific areas in the lower level of the U/ICC will be dedicated to this. Gravity will not be an issue for those coming from similar environments but certainly for those whose own home worlds are considerably different from Earth. In most cases, the latter will have undergone certain acclimatizing before arriving. Those coming in interdimensionally will most likely adjust with greater ease as, has already been stated, they will have learned how to shift and transmute matter and can calibrate their own bodies to a vibration suitable for Earth.[31]

<>

Keep in mind, coming to Earth for certain species will be easier for some than with others. In particular, this is true for those from dimensions of higher frequencies who will simply *see* themselves here and will *be* here in the same instant. Such species more often travel not with *spatial coordinates* but by a combination of *thought* and *feeling*, the result being sudden materialization and appearance. This has already been happening on Earth for centuries. The difference now is that in the time of the actualization of the U/ICC as seen in the *recurring dream*, the human species will have evolved to a point of interacting with deeper awareness and care. This is opposed to predominantly *reacting* to encounters from star visitors with fear, as has been the tendency throughout Earth's ages.

---

[31] Appreciation for adapting to gravity differences occurred for humans when preparing for moon landings.

- **The Center of Top Level:**

At the onset, the very top level of the Center building of the U/ICC, will also be dedicated to interaction with our star visitors. Details of this will be worked out later, remembering that the levels between the very top and underground will be dedicated to human development, particularly in awakening the heart energy within the self and the collective. Here is where the young children and elders play a significant role.[32]

In this area of the top level of the Center building, from time to time children and elders may also play a part as *"early greeters"* to those visiting from other systems. This is significant as the innocence of the children and their ability to experience *wonder without fear*, supported with the wisdom of the elders, will create helpful and healthy bridges for more sophisticated communication between humans and visitors.

The *shape* of the top center level will be dome shaped, a hemispherical rounded ceiling. This shape will work to enhance and distribute the pyramidal energy formed by the crystals embedded throughout the U/ICC. The substance of which this ceiling will be made will work in conjunction with the shape itself. (Others who've had this *dream* will have this information.)

---

[32] More on the role of children and elders is also described in Chapter 10 and later in this Chapter.

The shape and location of this area of the U/ICC will also support the eventual existence of *space-time portals* necessary for certain kinds of intergalactic and interdimensional travel.

- **Center Building's Role in Energy Flow for Entire Facility:**

    Imagine a *prana tube*.[33] This, in effect, is what is being created in the very center of the facility. Through the *prana tube*, energies of the Earth can flow upward and energy from the cosmos can flow down and into the "tube." This will energize the entire facility and all within it. The crystals within and surrounding the facility will enhance the effect of the *prana tube*, creating a bi-polar magnetic field, sometimes called a *tube torus*. (See Illustration #8, p.174.) This entire concept and the result will help to balance the energy of the U/ICC. This will be ongoing if for no other reason than the physical design of the structure itself produces this result.

- **Sound:**[34]

    > *Programmed Sounds:* As *sound* is able to activate specific states of consciousness, its study

---

[33] The human body is a bi-polar magnet with a central column called a *pranic or prana tube* or *antahkarana* by some ancient traditions, through which the life force or *prana* moves. The energy tube itself is open at both ends and runs from the crown of one's head down to the perineum. Also see Chapter 7.

[34] Also see Chapter 14 on *Sound*.

within the U/ICC will be brought into the *awareness training* through exercises and classes of experiential learning. Programmed sounds will be studied and put into use not to manipulate but to *free* the mind and will, and in this process, to show what is possible. It will be created in a way that the highest vibration in all life within and without the facility is nurtured.

➢ *Interdimensional Travel*: As you already know, the study of *Sound* will be essential in the U/ICC curriculum. Humans will be taught to remember what they have forgotten en masse within the collective consciousness that by shifting the vibrations of the physical matter of the human body, the frequency changes. This allows for a variety of things to happen, not the least of which is opening the door to having a different and new perspective in all areas of one's life. Interestingly, it also opens the possibility of interdimensional travel. Travel into *subspace* or *different dimensions*. This is how some visitors have come and will continue coming to Earth. It is also how human beings eventually will go to other worlds not located in the *physical* Universe.

- **Building Materials:**

    There will be *materials* used in the building of the U/ICC not yet discovered [35] but will be found in the very near future. This particularly has to do with *metals* that will be used. Certain combinations of molecules will yield substances capable of enhancing the *communicative qualities* of the *crystals* embedded in the foundations and walls of the overall structure itself.

- **Angles:**

    The *angles* of the building complex seen in the *dream* are characteristically *feminine* with flowing curves and few sharp angles. This creates an atmosphere for *allowing* a natural flow of energy and interaction, whether or not one is conscious of the experience. It is an environment that stimulates creativity. This design is much preferred over sharp prohibitive angles into which energy can be trapped or cut off. Why is this important?

    A fundamental purpose of the U/ICC deals with *communication*. Thus, rounded angles provide a setting in which the whole experience of communication is encouraged to roll naturally, with little constriction, not getting caught in perpendicular intersections (corners.) The energy will be conducive to empowering, encouraging, supporting and listening,

---

[35] At least such materials have not been discovered at the time of this book's release.

that all might hear and be heard. This vibration will be a result of the building structure itself, a vibration instilled into its very design. This is true for all aspects of the U/ICC, including all buildings and Gardens as well as Bridges.

- **Crystals:** [36]

    Crystals work in a variety of ways. It is necessary first, to state that crystals are *sentient*, a fact that is not yet acknowledged or even known by most humans, given the present state of consciousness of humankind. There are a few on Earth, however, who do know the nature of crystals.[37] Some of these individuals will be included in the design process of the part of the U/ICC that involves crystals.

    Crystals, since the beginning of Earth's formation, have been helping the Earth and other elements to hold and maintain the form of the planet itself, all waiting for a time when the *vibration* shifts for everyone and everything. That time has arrived.

    Therefore, in the context of the U/ICC, crystals will contribute and assist in two vital ways:

---

[36] Also see Chapters 14 and 15 on role of *Crystals*.
[37] It is worth reading Dhyani Ywahoo's book, *Voices of Our Ancestors*, in which she describes her role and that of the Ywahoo Clan of the Cherokee in caretaking the crystals of earth, believing themselves to have Pleiadian roots.

1) They are key to sending light or energy beacons off-planet, signaling visitors of other systems of humankind's state of development and stage of awareness.

2) They will help to hold and maintain both form and vibratory level of the U/ICC.

There is a conscious effort here to assist in helping the human species move into its next and higher state of evolution.

- **Streams of Water:**

    Inside the entire complex itself there are *moving creeks or streams of water*. This is to help keep the energy balanced and moving. The water works with the crystalline energy.

- **Plant Kingdom:**

There are plants in existence now whose stage of evolution allow for *doorways* to be opened to a clearer understanding of the interconnectedness of all life. Some might use the term "Oneness." Such plants (trees, flowers, bushes, etc.) will play an important part in the Gardens of the U/ICC.

> [*Author's note:* I'm reminded of an experience many years ago when I was attending an event in California. I lived in Tennessee at the time,

two and a half thousand miles away. At some point during a meditation, I was suddenly aware of a particular tree that grew in my yard at home. The sensation was one of the *tree's consciousness* being in direct contact with my own. Not only was this tree somehow "communicating" with me, it seemed to be the *spokesperson* for the whole forest of trees surrounding my home back in Tennessee. This was an incredible experience for me. Particularly when I learned there was a massive forest fire that had just broken out in the hills outside the city where the conference in California was being held.]

In addition to those plants involved with the Gardens, certain plants will grow and flourish inside the physical buildings themselves.

As to human awareness of plant consciousness, Indigenous peoples on Earth have always kept alive their connection to both the plant and animal kingdoms. And now there are a number of ecovillages on Earth in which the consciousness of trees and other plant life is central to the purpose of such endeavors. Findhorn in Scotland is one such example.

- **Animal Kingdom:**

    The creatures that live on Earth are not blind to the

shift in frequencies occurring at this time. They are alert and aware, able to move in harmony with all around them including human beings, in so far as humans are not harming them or destroying each other.

> [*Author's note*: I recently dreamed of two panda bears being encouraged to fight as entertainment for humans. I could sense other animals in the background being encouraged to do the same. Rather than fight, the two pandas looked at each other and both chose to lie down and die. This was the end of the dream.]

There will be those of the animal kingdom who will become an intimate part of the environment of the U/ICC, many of whom will roam the grounds freely. This atmosphere will help to sustain balance in the human psyche, encouraging deeper communication between humans and other lifeforms on Earth as well as the planet Earth itself.

- **Balance:**

The *crystals, water* and *plants*, in particular, will work together to create *balance*. To purify and move energy. This will be soothing to those humans undergoing more severe changes in their *awakening* process which will occur mainly in the beginning stages of the U/ICC. This will also create an environment conducive to open healthy *communication* between humans themselves and between humans and all lifeforms, on and off-planet.

As everything in the overall facility will be set up to create *balance,* this becomes the core perspective around which the surrounding community *(Universe City)* will grow. It will be important to always be willing to see with different, new, spherical, and *open eyes*, able to see at a three-hundred sixty degree scope both inside and outside the facility and its surroundings, as well as inside/outside the self. There is no difference, one being a reflection of the other.

- **Tunnels:**

There will be *tunnels* involved in the connecting of the *underground or underwater* areas of the U/ICC. These tunnels are part of the crystalline Bridge network and will have multiple functions, the first of which is to connect one part of the overall facility to another. They, too, will serve as *enhancers* of the energy that will flow throughout. This latter will occur as a result of their design and of the composition of material used to build them.

For those particularly sensitive to energy and in gauging the capacity of flow, the detection here will be obvious and easily measured. Eventually, over time, everyone will be able to sense and measure the "energy dance" that will occur once all is in place.

- **Transit System:**

    [*Author's note*: Information on a *Transportation System* showed up more than once over the past decades, including a *second recurring dream*. Because of this and the new information received during the time of writing this book, I have dedicated a separate Chapter (#18) to this subject.]

- **Extrasensory Skills:**

    Such areas as *telepathy* and other known extrasensory abilities that lie within all humans but which, for the most part, are highly under-developed, will be pursued in a more expanded and intentional way. The science itself most likely will be studied in the *East (mental)* arm of the U/ICC. Then the practice and implementing of such skills will happen in *all* arms and levels of the U/ICC complex. This will include the development and honing of *Intuitive* skills.

- **Open Heart and Atlantis:**

    [*As has been stated earlier...*] The small children and elders, when at the U/ICC, will spend most of their time in the Center building of the overall structure. One of their greatest contributions and responsibilities is to hone the energy of the *open heart*. Everyone will be affected by this. All will find their own *heart energy* opening and balancing.

[*Author's Note:* Though it may not be understood in the beginning by some, this singular process of *opening the heart* is the fulcrum point of my *recurring dream*. In other words, it is the direction in which the University/ICC must go to fulfill its purpose.]

To assist with this, there will be powerful opportunities integrated into the curriculum of the U/ICC specifically designed to help a person move through whatever emotional journey necessary to arrive into an *open heart* space. This journey most often includes dealing with and resolving issues that involve fear, anger and other related emotions. *Experience says that accelerated evolution and growth lie in feelings. This is especially true in learning to move through fear to open up to the experience of love.*

This particular area of the U/ICC is one of the elements that was missing in *Atlantis* [38] and, indeed, was never acknowledged, honed and nurtured. Regardless of whether one believes such a place as *Atlantis* ever existed, *experience also tells us that failure to garden the heart leads to an ultimate chaos from which only demise is the result.*

---

[38] See Chapter 19 on *Love*.

- **More on Children:**

    The children will play an important part in so many ways. It is actually the children's wisdom and abilities that can significantly teach the adults in areas that are of a *psychic* and *intuitive* nature. The elders, who will spend most of their time with the young children, will help to create the exercises and games that will uncover, improve and sharpen these skills for all.

    Part of what children will also do is reawaken the *innocence* born inside all humans, helping humankind to *wake up, see and be attentive* to all around them.

- **More on Children's Games:**

    As a child grows and matures, *children's games* will be especially useful in laying down the fundamentals of all else that will be made available in the U/ICC. Such games will be configured to be beneficial in helping adults and interstellar visitors as well. Exposing a star visitor to the foundational essentials given a human child will present a unique opportunity to learn about the human species.

    > [*Author's Note:* Only a few days before completing this book I heard a report on National Public Radio (NPR) that caught my attention. A 4th grade teacher by the name of John Hunter was being interviewed having just

published a book called *World Peace and other 4ᵗʰ Grade Achievements*. What was amazing was he had created a game with his 4ᵗʰ graders that posed global problems which demanded solutions that politicians, military leaders, tribal leaders, bankers, and diplomats have not been able to find for decades or even centuries. The children, through the game, found solutions to problems that seemed to be unresolvable. This prompted Hunter's being invited to the Pentagon and into other strategic arenas to share his findings. What the children had produced to solve world issues was no less than astounding, bordering on miraculous!]

- **Consciousness:**

The whole subject of consciousness will be explored in much the same way as those at the Institute of Noetic Sciences have done.[39] Consciousness is more than a topic to be studied through philosophical and scientific dialogue as will occur in the *East (mental)* arm of the U/ICC. It will be explored and experienced throughout the entire facility.

---

[39] The Institute of Noetic Sciences (IONS), founded by Edgar Mitchell in 1973, is "an organization dedicated to expanding science beyond conventional paradigms," which includes the exploration of Consciousness. IONS headquarters are located near San Francisco, CA, USA.

The awareness that *all life is connected* and is indeed part of a single consciousness will be the natural result of the environment the U/ICC will create. Such awareness will be strengthened through the nature and essence of the Gardens in particular and the activities that will occur there.

Accessing the subtler spheres of consciousness will naturally be part of this area of exploration in the U/ICC. As well, consciousness will be actively studied and pursued to the point of the creation and recreation of Ideas and their manifestation into the physical realm. *Sound* will also assist in this process.

> [*Author's note:* In Chapter 14, I describe the role of *Sound* in the U/ICC, relating the specific function of the vowel sounds in combination with the *spontaneous* creation of sound or "wordless songs." I often describe the experience as being one of riding on a "magic carpet" that when ridden takes one back to the original *Sound/Vibration* from which everything comes. This is another way to say, "…going to the point of creation of Ideas and the perceptual realities that follow."]

- **Changing Universe:**

The Universe itself is changing. It is expanding. This is happening in the physical Universe and is occurring interdimensionally. Not one act happens in isolation. *One act affects the whole*. As what has been seen in the *recurring dream* begins to manifest, it will affect and change all. It has already begun.

- **Hundredth Monkey Effect:**

In the daily routine of the human being there will be a natural metamorphosis taking place. In a short period of time, the "hundredth monkey"[40] effect will be reached and the shift from "the shallow to the deep" will occur throughout the human species. This means the consciousness of humans will quickly adjust to the new vibration taking place. One confirming measure that will bring some peace to those of a scientific bent will be the increased use of broader human brain capacity. There have been decades of preparing for this. The human species has been receiving help for centuries from star beings, but only in the last thirty or so years has there been concentrated effort made by those of other systems to

---

[40] The Hundredth Monkey is an apparent phenomenon in which behavior or thought spreads rapidly from one group to all related groups once a critical number of participants is reached. Much of this theory is based on the observation by scientists of the behavior of the Japanese monkey, Macaca Fuscata, over a period of thirty years. When one monkey began to wash a sweet potato in the salt water of the sea, other monkeys of the same tribe on the same island began to follow suit. Suddenly, monkeys on other islands began to wash their potatoes in the sea. At some point it was concluded that the energy ratio increased to such a level that "washing potatoes in the sea" entered into the monkeys' mass consciousness and became part of the monkeys' collective behavior.

nurture and prepare humankind for the shifts here now and those coming.

- **Vibration Increase:**

   Inside the structure of the U/ICC itself will be what later will be perceived as *elementary exercises* and *practices*. These are to better prepare the human consciousness to accept the variance in vibratory levels that will soon become the accepted norm. As time continues in a linear fashion, the vibratory levels will continue to increase. This will happen naturally and enough will have been put into place by then that humans will adjust more easily. Also, the human consciousness will have collectively moved into a state of *awareness* that is able to hold what before was perceived as an impossibility.

- **Conducting Garments:**

   There will be garments made available in the U/ICC whose very substance acts as a conductor and enhancer of energy which assists humans and star visitors alike in finding balance within their own energy fields. This will be especially beneficial for those coming from other systems having to adjust to conditions on Earth, physical and otherwise. This includes adjusting to the energy field of emotions (which appears to be lacking in certain species and may be a new experience for some.)

- **Faraday Spaces and Silence:**

    There will be *Faraday* rooms and spaces made available for humans and for interstellar visitors.

    > [*Author's note:* For those who may not be familiar with a *Faraday room* or *cage*, it is an enclosure that blocks external static and non-static electric fields. Though it doesn't seem to block the Earth's magnetic field, it can shield from external electromagnetic radiation and create an environment free of electromagnetic interference. My first experience with a *Faraday cage* was in the home of a friend, Dr. Edith Jurka.[41] Living north of New York City in the village of Croton-on-Hudson, Edith had created a *Faraday cage* out of one of the rooms in her house. Walking into this room lined with copper and other conductive materials and sealing it by closing the thick heavy door made of the same materials, I experienced a depth of silence that was unlike anything I'd encountered before. The silence was deep. The stillness, profound.]

---

[41] Edith received her medical degree in psychiatry from the Yale University School of Medicine. She was highly respected among her peers, and was well known for her sense of adventure in the area of science and her willingness to move "outside of the proverbial academic box."

Faraday areas will allow *breathing spaces* for all species, human and otherwise, to adapt to the changes necessary to be in a higher vibrational environment. For humans, in particular, such areas will allow for an easier smoother shift into more advanced states of awareness as the *Faraday* room gives complete and total silence to the mind and kinetic activity that is otherwise in constant motion, motion that is stimulated from "outside" electrical and magnetic influences. *Silence* is the goal here. From the *silence* comes the order and wisdom that encourages an ongoing *peace* and *centeredness* within the individual and collective self. Do not underestimate the importance and necessity of *silence* and the *stillness* that accompanies it.

- **Conversation Rooms and Active Listening**

There will be *conversation* rooms into which specific species, both human and off-planet visitors, may go to interact. As some may not initially appear in physical form, the silence and intention will set the "stage" for interaction to occur between some species.

- *Active Listening* will be taught and honed as a necessary skill for effective communication, though the physical exercise of *listening* will reach far beyond that of the ears. It will become a multileveled skill, embracing considerably more than what is involved in the act of physical *hearing*. This subject will be part of the curriculum at the U/ICC and will be required to be learned by all as *listening* is fundamental to successful communication. Already there is much in this area being taught and learned.

As humans move into the future years, it will be necessary for the eventual survival of humankind to learn *deep listening* to the point of its becoming inherent to near instinctual. Simultaneously this skill will become intrinsically central for healthy communication to transpire between species. *Listen. Listen. Listen.*

- **Lighting:**

The lighting inside the entire facility of the U/ICC will start off in a traditional way at first but will graduate into a different source of energy, one that is environmentally safe and even nurturing to those who use it. Lights will emit rays that interact with a higher vibration in a way that an *energy dance* occurs. In other words, the system of lighting itself will be productively feeding and interacting with primary energy sources, most of which will be solar driven.

In time, however, there will be new and different sources that come from tapping into the *life force* that is in all things by using the energy of one's mind or will. Mechanical devices, for example, that switch lights on or off, will not be needed. Already *touch* and *sound* sensitive devices that trigger electricity flow are becoming the bridge to a future where light and energy sources will be accessed merely through *thought* or *willing* it to happen. There are those (human and extraterrestrials) who are able to do this now, though this skill will not develop fully in the collective human population for a period of generations.

- **Breathable Air:**

    The air that will be breathed in the U/ICC will first move through a specially designed filtering purification system. This will be essential for both human beings and for star visitors, necessary primarily for the latter's adapting safely to Earth's atmosphere.

- **Colors and Patterns:**

    It is already known that frequencies of various colors can have an effect on the emotions and physical wellbeing of an individual. Throughout the total structure of the U/ICC there will be specific colors and patterns used to stimulate, enhance and nurture the *awakening process* of a person. The vibrant colors of the rainbow, for example, play a very real part in sustaining a *vibration* of continued *wakefulness*.

    > [*Author's Note:* When I had cancer I used unconventional methods to help myself heal, such as *sound* and *imaging*. One day a friend

asked what *color* or *colors* represented for me absolute health. Without hesitation I said *"blue magenta."* I was referring to two *sheer* pieces of material I had been gifted some years before and had kept around because of how they made me feel, not to mention that they were absolutely beautiful. One was royal blue in color, the other a fluorescent magenta. When placed on top of the other a color and effect was produced of such beauty that at times I would feel a slight shivering sensation that ultimately seemed to produce balance in my body resulting in calm and peace of mind. At my friend's suggestion, I draped pieces of various sizes of the two combined materials in strategic points in my home where I would see them throughout the day. The impact was amazing. Visually and vibrationally, these combined colors embraced me when I would see them. Always then I would feel a quiet stillness come over me… a *centeredness*. For me, there is no doubt that these colors played a critical role in my healing.]

- **Places for Nourishment (Cafeterias, etc.):**

As to places for physical nourishment such as cafeterias, this will look different for different species as food for one may not be experienced as a nourishment source for another. However, there will be common gathering places established for congenial relaxed conversation to occur.[42]

- **<u>Bathrooms (Accommodations for disposing of *waste* material):</u>**

Bathrooms for some visitors will not be necessary as their own energy bodies will transmute what might need digestion. Such visitors then will assimilate the energy that comes from the breakdown of their "food" sources directly into their bodies to be used as a fuel resource or they will send the energy transference *out* from their bodies, to revitalize life around it, much like what occurs when human body waste goes into the ground to act as a fertilizer.

For others, the engineering of specially designed apparel[43] and bathrooms will be created and adapted to accommodate various body forms. Remember that many of those visiting will come from *subspace*, i.e. a non-physical plane rather than a physical plane first. This will require some natural readjustment. Even

---

[42] Imaginative and colorful examples to reference are bar scenes out of the movie series *Star Wars,* and the upper deck on the starship Enterprise in the *Star Trek: Second Generation* television and movie series.

[43] A related scenario that comes to mind is that of recycling the water (urine) of one's physical body described in Frank Herbert's book *Dune* (later made into the movie *Dune*) in which the desert dwelling people designed a body suit that would allow the recycling of their own body water to help them survive the extreme conditions on their planet of Dune (Arakis.) The design and function of similar apparel could be combined with the conducting garments mentioned earlier in this Chapter.

though all coming to this planet will have undergone certain *conditioning* in order to come here, *being* on Earth and *preparation for coming* here are two different things, and could be challenging for one who has never been here before.

- **Simultaneous Learning:**

   As humans begin to adapt to the environment of the U/ICC and move further into its offerings and purpose, many levels of training will begin to occur *simultaneously.* In the beginning all will start out in a *linear* fashion as this is presently the perception of reality most comfortable and familiar to human beings. But shortly after the training has begun, several layers of experience will happen concurrently. An example might be the honing of telepathic skills as a person is also consciously engaging in a philosophical conversation with those around him while interacting with the consciousness of plant life in the facility, all occurring while carrying on a physical activity of some sort, etc. and so on.

> [*Author's note:* Deepak Chopra in his book *Creating Affluence* speaks of the organizing power of the universe, how all happens at the same time and in correlation with the other: movement of galaxies and stars, the rotation of the

earth, the cycles of seasons, biological rhythms of our bodies and everything in nature, birds migrating in a precise season to a precise place, fish returning to original spawning grounds, etc. etc.., again all occurring simultaneously and correlated with each other.

Then he speaks of the amazing ability of the human body to "think thoughts, play a piano, sing a song, digest food, eliminate toxins and germs, monitor the movement of stars, and make a new baby all at the same time, and correlate each of these activities with every other activity." He says that inherent in the field of all possibilities is this amazing infinite organizing power. *What the U/ICC will do is bring much of what appears to be unconscious action into conscious awareness.* The result of this will be an undeniable experience of the interconnectedness of one's own nature.]

As this kind of learning takes place, *spatial perception* will begin to shift. In the first stages of the U/ICC, children will perceive an expansion in their individual sense of time and space more easily than most adults. Children then will grow into adults comfortable in this way of being because they will have lived inside this new reality all of their lives. It will take two to three generations for this to happen but it *will* happen. As this occurs, the r realization of shifting in and out of linear and non-linear environments will become easier and this opens the door to experiencing planes well beyond the physical. In time, humans will be able to move through *subspace* as easily as many of their visitor friends.[44]

- **Twenty-Four Hour Curriculum:**

    There will be a constant twenty-four hour offering of subjects that will be determined by the overall community, as the purpose and integration of the U/ICC begin to root into the consciousness of the community itself. For example, there will be constant availability of experiential exercises to sharpen extra-sensory skills. There also will be experiences offered in music/sound. In addition to the curriculum itself, areas such as *meditation nooks* for individuals as well as groups will be made available.[45]

- *Night Time Curriculum:* Activities that involve the night time will be important, not only for learning about the stars, but also because *the energy of the night brings a balance to the human psyche,* allowing knowledge and other info buried inside the unconscious to surface. This stimulates

---

[44] Some indigenous peoples have lived in this way since the beginning of their time on earth. What is being spoken of now is bringing this awareness, ability and willingness to live in this way into the collective consciousness of all humankind.
[45] Also see Chapter 10, section on *Bridges*.

evolution of the human species. And when needed moves it at an accelerated rate and always toward a higher state of awareness.

> Certain garden activities will be a part of the night curriculum, particularly in regard to plants and animals of various species. Here again, this is part of helping humans remember their direct connection to all life. In the beginning all will be kept quite simple so that none of the "building blocks" are lost.

- **Relationship to Water, Land, Air:**

Not all of the communities/Universe Cities will have immediate access to sea waters but all will have access to land underneath which some water reserves will serve to support. Humans will learn how to create and access moisture from the air in more sophisticated and reliable ways than is available in the present day (of this book's writing). There will grow to be a *symbiotic relationship* between humans and Earth's atmosphere, unlike anything known in human history. As time goes on and more and more humans *awake*, there will come the ideas of how to "dance" in balance with all life on Earth.[46] The U/ICC will be a central element in helping the mass consciousness shift to make all this possible.

---

[46] Again, in so many ways this has already been occurring with indigenous people around the world. The intent now is to bring this state of awareness into the Collective Consciousness of all humans.

- **Gardening:**

   All will be taught to *garden* both *physically* and *metaphorically*. *Physically,* the process of tilling the soil, planting seeds, experiencing their growth and development, experimenting with plant life communication, these and other possibilities of plant interaction will produce results that will nurture the entire community. *Metaphorically,* each physical step taken will have its counterpart in the world of symbolism, and bringing this to light at every turn will act as a catalyst to awaken the mass consciousness to the fact that all life is part of a single whole. There is much to be learned here. That the human species as a collective experience direct connection to all life on Earth and beyond Earth, is imperative. [*Also see Section on Plant Kingdom in this Chapter, pp. 120-121*]

- **The Five Elements:**

   It is not by accident that many of Earth's cultures recognize *four primary Elements (fire, earth, water, air)* as the ingredients that make up *matter,* and which in combination create what is known as the physical world. There is a *fifth Element* that is recognized by some and said to be beyond the material world. The ancient Greeks called this *Aether* (or *Ether.*)

Awareness of the Elements will become an intrinsic part of the U/ICC culture. This is important as humans come to understand and appreciate the living consciousness that lies within each Element. Much of this wisdom will become accessible through the work and production of certain *sounds.*

> [*Author's note:* Over the years I have come to have a deep appreciation of the different scientific and philosophical perspectives regarding the nature of our physical world. I have also learned to listen to my "gut instinct" (intuition) when presented with a concept that requires a modicum of discernment on my part. Granted there are some people out there offering some wild and seemingly impossible ideas and viewpoints, and some of you reading this book might include me in this prestigious group which I would readily see as a compliment. I do, however, want to share a concept regarding the *Elements*, specifically *earth, fire, water* and *air*, that inspired me to "move outside the box" and explore a possibility that I discovered my "gut" already surmised was true. One might call the following a "poetic" perspective on the nature of the physical world. And for me a most viable perspective. To quote the Hathors in Tom Kenyon's book*:*

> *"What has happened at this plane of existence is truly a miracle. For these four beings, these vast beings of Earth, Fire, Water, and Air/Space, have joined together to allow you to have a physical bodily form. It is a gift that is given freely so you might have the benefit of experience in a denser world than from where you come. Without these conscious beings' work and mutual cooperation, under the creative desire for your existence here, there would be no evolution possible on this three-dimensional plane. In fact, there wouldn't be a physical plane at all. "*]

The concept that the *Element*s are conscious beings who have intentionally come together to allow the experience of the physical world may push the boundaries for some, yet if seen through the innocent eyes of a child, has worth and surprising possibility.

Consider also that the *Element*s are an "archetypal dimensional blueprint for the unfolding of consciousness into matter ..."

Where all this fits into the purpose of the U/ICC lies in the first steps that will be taken in helping human beings to awaken to the principal make-up of themselves and their world. This awareness will then open a door in the human psyche allowing for further and more complex understanding to enter. All of this will be a necessary focus of the curriculum of the U/ICC.]

- **Vibration of Love:**

    [*Author's note:* Because the topic of *Love* is so vast in itself, to attempt to describe it's intimate role in the life of the U/ICC borders on the impossible. Therefore, recognizing the enormity of the subject but also realizing the necessity of sharing what I've learned in the transmissions and have come to know regarding the fiber and character of the U/ICC, I've chosen to devote a full Chapter to *Love.* (Chapter #19)]

- **Monitoring Council:**

There will be a body of individuals comprising the equivalent of a *Think Tank,* with expanded purpose to it. The group itself will be made up initially *of thirteen* people, *three* of whom are *children* (12 years or younger), *three* of whom are *teenagers* (13 years plus), *three* that are *elder*s (sixty years and up), and the remaining *four* being <u>adults</u> (generally between 20 and 60 years of age.) In time, a single representative from visiting species will be invited to be a part.

This group will be the *monitoring body* of the U/ICC. It will be the known but *unseen* nerve center, monitoring the health of the U/ICC, seeing that its high purpose is kept intact, that at all times the focus remains on the two parts of the U/ICC's purpose: 1) raising the level of human *mass* consciousness to a full and perpetual wakeful state with awareness of the interconnectedness of all life; and 2) cultivating healthy communication and interaction with other species not of planet Earth.

An added responsibility and a natural outgrowth of this group will be to capture the creative juices that will flow when the group comes together. In other words, once recognizing that the U/ICC when coming into being has within it all the characteristics of a *living organism*, and accepting the fact that all life moves through certain stages of growth, this monitoring group, being an intimate component of the U/ICC's body and lifeforce, will *naturally create new and innovative ways* to support and enhance the healthy evolution of the U/ICC. (This intimate interplay between the *Monitoring Council* and the U/ICC will determine the appropriate time of service *needed* by those within this *Council* to fulfill the responsibility they've accepted. Once determined, the *length of service* for an individual *council member* will be ultimately decided by the larger community.)

- <u>*Children's Council:*</u> Consider the creation of a *Children's Council* as a subset to the more eclectic Monitoring Council. The unbridled imagination of children coupled with their innocence and curiosity, brings a wisdom that a *transforming mass consciousness* needs as all move into a new way of being.

<><><>

CHAPTER 18

Second Recurring Dream: A Transit System

This book is dedicated to bringing to light and life a specific *recurring dream* that occurred over a wide expanse of many years. What I haven't spoken of is a *second recurring dream* that also came in my earlier years, long before any of the science fiction depictions of the future crossed my path. This is a dream I actually had forgotten about until the *transmissions* started coming. It was only then that I realized that the *first dream* (of the U/ICC) and this second dream are connected.

And so, what is this other dream? It is of a *transit system*.

Here is the dream …

> *I'm standing in what looks like a very wide "tunnel way." It seems to be negotiating distances underground or underwater, connecting one point to another. Though it is under the Earth's surface, the lighting is bright and simulates sunlight. The architecture is very sleek and advanced. There are "moving walkways" calibrated at*

different speeds, one speed graduating gently into another, allowing a person to choose which walkway better suits the desire of the pedestrian. All walking tracks are color coded designating the rate of speed at which one moves.

In addition to the moving walkways there are "transparent tubes" that appear to hold up to four passengers each. These tubes move in a trainlike fashion allowing passengers to sit and move greater distances at instant speed.

The reason for such a transit system came clear in the *transmissions* and helped me to see why the two *recurring dreams* are linked. Such a *transit system* is to connect the surrounding communities to the U/ICC which is placed at the heart of the Universe City. There will be no "parking lots" that lie in the immediate area of the U/ICC. Rather those bringing their own vehicles (whatever the technology at the time... Air cars?) will park them in provided areas off the mountain or hill upon which the U/ICC is built. At that point the people can access the moving walkways and/or tubes primarily designed to carry people from distances not in the immediate vicinity, that then will take them to the arrival areas of the U/ICC.

One piece to include here is this. In the beginning all passengers will be human. In time that will change.

I do not have details on this and how it will all transpire but I do know that there is a system that will be put into place that will allow for visiting vehicles not of earth to land and be housed. Keep in mind there are those that will not need such facilities as they will arrive via portals, or some such, within the U/ICC itself.

Taking a step back a few feet to observe, it isn't hard to imagine a gigantic puzzle forming, one whose pieces are composed of the *transit system* and each and every transmission and dream that has found its way into this book. And it may be that one of you has dreamed a piece of this puzzle as well. Once each piece finds its place in the overall picture, the final result will be one of indescribable magnificence and limitless possibility.

PART IV

Moving Forward

*"In terms of human evolution,
Love is where you're heading."*

- *The Hathor Material*

CHAPTER 19

Love

The concepts of *love* and compassion haven't been mentioned very often in this book. Yet, as stated in Chapter six, *love* is the underlying premise of my *recurring dream* and the founding of the U/ICC and all that surrounds it.

Two quotes worth repeating here are these:

> *"Love and Compassion are necessities. Not luxuries. Without them humanity cannot survive."*
> - Dalai Lama

> *"The highest state of consciousness through which you can ascertain the truth of a situation, Is through the vibrational field that you call Love."*
> - Tom Kenyon,
> The Hathor Material

I remember years ago when my husband John and I were in Israel. We met an extraordinary rabbi who spoke of nothing but *love* non-stop for nearly an hour.

John made the comment later that he had never met anyone who could describe *love* in as many ways as this rabbi did. Over the years I became more and more aware of the impact the vibration of the word itself can have on a situation, regardless of the language in which it is said. I've learned that *love* is more than a feeling, more than a philosophical concept to be considered, more than almost any description that can be assigned to it.

I have come to the full realization that *love* is the essence of all life. It is the Source from which all else originates, whether it is an idea or the manifestation of that idea. I've heard it described as being the "stuff of which the Universe is made."

I have also come to the understanding that responding to life from a place of *love* elevates a person's level of consciousness to one of clarity and discernment not accessible otherwise. This is crucial and becoming more and more apparent as humankind explores new and further indepth interaction with itself and with other species, on and off-planet. In fact, this is the *crux* of the U/ICC's purpose.

I'm not going to attempt to philosophize here on the countless viewpoints, the pros and cons of various individual and collective perspectives on the subject of *love*. Only to share that *love* is key to the manifestation and ultimately the character and fiber of the U/ICC itself.

It was when I read Lynn McTaggart's *The Field* that I had an "aha moment" reading about Rupert Sheldrake's "morphic fields." Sheldrake describes what he calls a field of *morphic resonance* in which there are self-organizing properties of biological systems, from "molecules to bodies to societies," and that these are fields that " influence like upon like through space and time." In layman's terms, "the more we learn, the easier it is for others to follow in our footsteps." I speak about this in the *Hundredth Monkey Effect* in Chapter 17. Where this directly applies to the U/ICC and what has prompted the "aha moment" is this: Creating a "field of love" as the basis upon which my *recurring dream* is built ensures a sustainable environment for the ongoing higher development of consciousness. That as generations come and go, the essence of the *divine* and sacredness of all life is nurtured and encouraged.

A piece of information that came through in the *transmissions* that I'm choosing to include here that specifically has to do with *love* is this:

> "There are unlimited forms of creation. This includes all that is familiar to Humankind, plus forms of creation not yet discovered or

encountered. *All life responds to love.* All. *Love* being the highest known vibration to exist. As star visitors and human beings come into contact with each other, the vibration in which each moves and lives at the time will define the 'playing field' of interaction that will take place between species. Even those that at the onset who appear aggressive and 'militant' will respond to the vibration of *love.*

"For those species *choosing* to remain in a lower vibratory frequency, anything and anyone that is moving at a higher vibration will move from their field of recognition. In other words, those beings of a higher vibration will seem to disappear.[47] They literally will be 'out of sight' to those who are existing at a lower frequency. This phenomenon will occur until a time when the mass consciousness of that lifeform raises in vibration, whether by conscious choice or natural evolution. In the U/ICC, the vibration will be kept *high*. This will occur not only through the physical manifestation of what has been seen in the *recurring dream*, but specifically through the attitude and spirit of those accepting the responsibility of bringing a shift

---

[47] There are documented accounts of persons claiming to have close encounters with extraterrestrials who speak of this phenomenon, stating that those beings moving at a higher frequency seem to disappear unless these star visitors *allow* themselves to be seen. John Mack describes some of these encounters in his book *Passport to the Cosmos.*

> in human consciousness into being. Any lifeforms, however they appear in size, shape or substance, will encounter a vibration that is one of *Love*. "

At this point, it may be important to share something I learned about the supposed fate of the legendary civilization of Atlantis, something to remember as we move forward in bringing the U/ICC into being…

There have been many stories and theories about the last days of Atlantis, first mentioned by Plato in his Atlantis Dialogues written around 360 BC. Plato spoke of an ancient city that existed in approximately 9,600 BC, whose inhabitants were said to be blessed with abundance in every way, materially certainly, as well as in technological and scientific accomplishments and mastery. And with all this, they were particularly blessed with a *divine nature* and *wisdom* that guided them in every aspect of daily living, including interaction with each other and their environment.

As long as the Atlanteans continued to honor the "*divine nature"* within themselves and each other, the abundance of blessings in their lives continued to grow and increase. It was when the "*divine portion*" of their nature began to be neglected that the perceived utopia of Atlantis began to go off track and lose its way. Many at the time who were immersed in the idea of abundance, however, were blind to the fact that their civilization was heading down a road toward extinction. The theories that speak of Atlantis's final days include scenarios of volcanic eruptions and earthquakes. Whatever the facts, the *truth* of the metaphor that comes when the *divine nature* within humans is negated still stands. When humans begin to forget to *love* and nurture each other and their environment, they begin to lose their way. And then, regardless of the sophistication of technological development and the intoxication that can comes from acquiring abundance, if *love* (in all its many faces) is not fundamental to the foundation of a civilization, then its downfall is inevitable.

To reiterate a thought shared earlier: *Failure to garden the heart leads to an ultimate chaos from which only demise is the result.* An added perspective is to consider the *heart* of a civilization, or in the case of my *recurring dream,* the *heart* of the U/ICC. To garden the heart means to come from *love* at all times. Simple but not always easy. For the U/ICC, absolutely necessary.

How all this will manifest, I can't yet say. What I can say was given in the last paragraph of the Manifesto I presented years ago:

"As we move forward into the manifestation of this dream, the beacon of light that guides us lies in the following questions: 'How are the people treating each other as together we create this experience?' 'Are we being kind with each other?' 'Are we listening?' 'Are we able to be in integrity with ourselves and others and do it in a caring way?' These are the questions that will guide those of us who are called to be a part of this *Vision* ... this *Dream* ... "

And so ... "University and Interstellar Communication Center??" ... "U/ICC??"

If one listens to the name itself, the message becomes loud and clear ..."You and I *see*." OR to borrow from the movie *Avatar*, "I *see* you." Either way, the message to each other and to our friends from near and distant galaxies is, "It's time. We're ready."

CHAPTER 20

Paradigm Shift

What I have seen and proposed in this book, as I have already said, borders on *science fiction* and certainly has all the characteristics of the *impossible*. What then is necessary for a *dream* of such magnitude to become a full blown physical reality??

First, it takes more than the *belief* that the vision seen in my *dream* can be manifested. It takes the *knowing* that such a vision is doable. And here's the thing. *It only takes one person who is willing to say "yes."* This then puts a hairline crack into an existing belief system that no longer serves the good of all.

Once this occurs then the laws of nature take over and in time, the crack widens to such a degree that that entire belief structure shatters. When this happens, a door is flung wide open for a *shift in consciousness* to occur.

The question then becomes, "Who is going to walk through?" Already there are countless thousands, maybe millions, of people *waking up,* becoming aware that something has got to change. Change in the way we think, the way we do things … the way we treat each other. To continue down the path we're presently on is a recipe for no less than disaster for the human species, if not for all life on Earth. Regrettably the message that gets repeated over and over again and that has ensnared so many for eons is this belief, "I am only one person. What can I do??"

Granted when one is caught inside a reality that communicates such beliefs as "It's no use. This is the way things have always been," OR "Forget it. There's no one listening," it is easy to understand why the masses throw in the towel and simply give up. This is the ideal formula for mass despair and depression which in turn is fodder for power hungry egos. Taken together these scenarios are the perfect catalysts needed to keep the concept and perception of *separation* in place.

Separation at this level begins with an individual's judgment of self that bleeds out into the judgment of community, local and global, touching all facets of one's reality. The ego, set on claiming identity for itself, creates situations in which the eventual outcome is nearly always detrimental to the emotional, physical, mental and spiritual health of everyone and everything with which it is interacting.

This we have seen and are continuing to see played out in every aspect of our daily lives, including the environments we've created at home, at work, in education and government, and so on. Ultimately this approach to life is neither viable nor sustainable. Given this, the crumbling of the foundation of both the individual and the collective becomes inevitable.

One of the most powerful and insightful books to come along is John Renesch's book, *The Great Growing Up*. In it he says, "We are becoming frightened, collectively addictive, personally isolated fundamentalists who have been avoiding growing up and committing ourselves to living together in harmony." This pretty well sums up the present state of things. AND …

Knowing this, taking place at this exact moment is a stirring inside the collective that is so uncomfortable that it is clear *change* at a scale that defies the imagination is imminent. Most people probably don't have a clue as to what they are seeing and experiencing in the changes already in motion, but I like to think there are people *waking up* at a pace that challenges the speed of light.

Fortunately, there are increasing numbers who are consciously choosing to create different realities for themselves and in doing so they are planting seeds that effect the whole.

I have often said, "It only takes one person to shift an entire Universe." The prospect that there is more than the *one* out there willing to say "yes" is what I'm counting on. In fact, in the beginning of *Eyes Open* I stated that the reason for my even writing this book is to find the others of you who have literally had the same *recurring dream* I have had. *I know beyond all doubt there are at least twelve of you.* Perhaps more. Each with pieces to this amazing puzzle. And the time has come to find each other. Why? Humanity's collective consciousness is now at the brink of having to make a choice. Either we change or disappear.

A possible explanation of what we on Earth may now be facing, is illustrated by Renesch's sharing of David Kyle's[48] description of *change* and *paradigm shift*. Kyle says that when a *paradigm shift* is pending, preparing to take an evolutionary leap, " … there is often a 'bunching up' of the species or particles or whatever is preparing to make the leap, and they collect around a threshold." He continues to say that, "Change of state for any species is that point where

---

[48] David Kyle and John Renesch both are consultants and faculty members of the Center for Leadership Studies founded by Mel Toomey, a pioneer in the field of advanced *leadership* programs and development. Kyle wrote and presented a paper to Yale University in the summer of 2005 on "causality consciousness."

the entire species faces an evolutionary jump toward a radical and unpredictable next step in evolution… Either the entire species jumps across the chasm together, or the species flames out and dies trying to change. This means the species doesn't *dribble* across a threshold. Either the whole group goes through the *transformation* or they die out together, as so many species have done over the history of this planet."

At this very moment in time, humankind is looking at the *chasm of change* in the face. I don't know about you but I'm making a conscious choice as one individual to be and do whatever is needed to put into place, a better world. I'm holding out my hand to take yours, trusting that together we can take a colossal leap into a future though unknown at the onset, is certain of success. Clearly, we are inside a birthing process of a new dimension and an expanded awareness of consciousness. What an extraordinary time to be alive!!

ILLUSTRATIONS

ILLUSTRATIONS
(Copyright 2012 © by Jeanne White Eagle)

Illustration #1 –

Sketch of Recurring Dream, Scene 1:

Illustration #2 –

Sketch of Recurring Dream, Scene 2:

Illustration #3 –

Thirteen Circles (Identified) of the U/ICC:

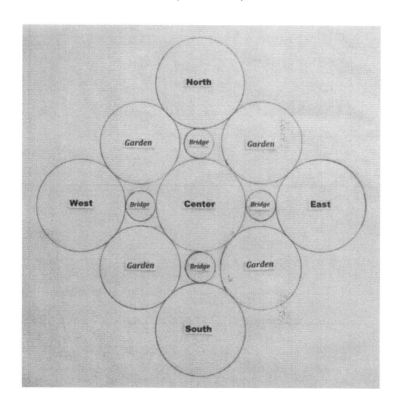

Illustration #4 –   Illustration #5

Golden Pyramids: Thirteen Circles:

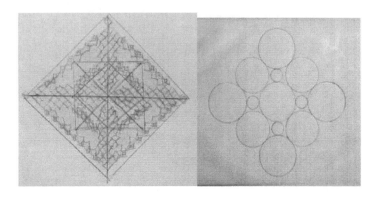

Illustration #6 –

Energy Pyramids on Thirteen Circles:

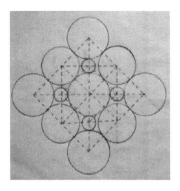

Illustration #7 –

Energy Beacons(Signals) resulting from pyramidal energy flow in U/ICC:

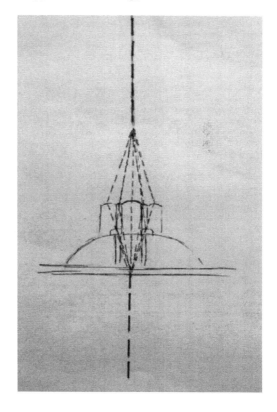

<u>Illustration #8</u> –

Two-dimensional Bi-Polar Magnetic Field (Prana Tube) created by the structural design and energy of the U/ICC:

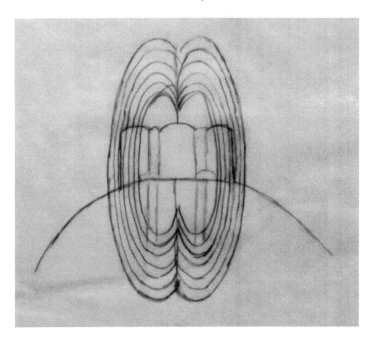

ACKNOWLEDGEMENTS

Writing this book has been both an adventure and a great joy. Not once have I sat down to write that I haven't experienced a sense of elation and excitement, somehow knowing that this *dream* that found its way into my psyche so many years ago has finally found its way onto a stage where others might share what I have seen. I suppose then, that one of the first shows of gratitude goes to whomever or whatever dropped the seed of this *recurring dream* onto my path decades ago. And with this I want to thank all the dreamers out there, those of you willing to stick your necks out and be called a "fool" but who nonetheless are brave enough to take those steps that give those of us, as myself, the courage to push the envelope to see what is possible.

In recent years since the *dream's* return, there have been a number of people who have stood by my side helping me to put voice to the vision. In particular I wish to thank *Sophia Peters* from Germany, *Hlin Hansen* and *Morten Østbye* from Norway who were my original "Board of Advisors" to be followed by *Tracey Turner Keyser*, *PaTi Coleman*, *Margarita DiVita* and *Candy Barbee* from here in the States , *Xochil Rubio* from Mexico, then *Anael Harpaz, Haghit Rosenberg, Irit Barber*,

Tami Semmel all from Israel. Other wonderful and supportive souls are *Cheryl Patterson(*TN), *Jo Fisher*(NC), *Wooddora-Rose Eisenhauer,*(CO), *Brett Almond*(UK), *Ocean Graham*(Scotland), *Beate Pruegner*(Germany), *Sammye Jo Harvey*(TN), *Rolf Schwarz*(Germany), *Florian Heins*(Germany), *Sonja Munz*(Bosnia), *Henry Rowan*(Ireland), *Susanne Fuhr*(Norway), *Chris Bruggeman*(Belgium), *Telse Mumm*(Germany), and *so many others* around the world whose names are too many to mention here but who know who you are. I hope all of you know how dear each of you is and how deeply appreciative I am to you for believing in me, more importantly for believing in the *dream* I've seen for a new world.

More thanks to you *Candy*, and you *Colby McLemore* for your patience and expertise in photography. And certainly you *Sean Pehrson* for creating one of the most beautiful book covers I've ever seen. You did an amazing job! No less than a million thank yous are in order!!

And of course, I can't forget *JJ* and those luminous off-world beings (Hathors, Pleiadians) who seemed to float in when most needed, including the birds who came daily to my window.

I particularly want to thank *John Renesch* for his patience and support in listening to my earlier thoughts that though full of enthusiasm were somewhat "discombobulated" in their presentation as I was still learning how to articulate the enormity of the vision I've been given. Thank you, too, to

Sue Mehrtens, who in her brilliance has listened to ideas that clearly did not fit into a traditional academic setting but who has supported and encouraged me all along the way. Thank you, *Herman Maynard* and all others who have read the manuscript before its publishing and who through friendship and comment have supported the amazing *dream* put forth in *Eyes Open*.

And thank you, *Edgar Mitchell*, for stepping out on a limb to write the Foreword for this book. You have been an inspiration in my life and in the lives of millions. May the days and years ahead confirm the steps we each have taken to bring about a better future for our children and our children's children.

A special thanks goes to *Ula Rae Mynatt* in whose country farm house I created my *writing lab* which served to be the oasis from which I drew strength and was able to put into writing the extraordinary ideas that have shown their faces in *Eyes Open*.

And always a special thanks to *Joe Nusbaum*. The wisdom of your guidance and constant support along with an ever present upbeat attitude has helped me to keep putting one foot in front of the other. You are amazing.

And then there are my mentors, those who came forward through varying circumstances to shine light on my path and to share their wisdom that I might find my way… *Kit Haynes* and *Perry Lane*, my incredible parents, *Herb* and

Jane Allen from Up With People, *Chick* and *Adrienne Davis*, *Willis Harman, Joseph Rael, Satya Saibaba* to name a cherished few. Thank you, my teachers.

And now I come to those who have given me the wind under my wings. To my Beloved Husband *John Pehrson*, forever thank you for being in my life. Thank you for your neverending patience and willingness to be there for me, particularly in those times when you had other things to do but always seemed to find time to hear me explore and test out passage after passage as I wrote to put my ideas into book form. You are closer than my breath. Thank you, too, for bringing *Sean, Alan* and *Ryan* into my life. I am richer because of it.

And to my wonderful and no less than amazing children, *Jenny Stiffler* and *John Troutner*, your ongoing trust and support continue to give me strength, particularly in times when I start questioning myself. There are no words to express my love and gratitude to each of you.

And finally I want to thank a precious and beautiful soul who has entered my life with grace and grandeur. That is my granddaughter, *Kendolynn Troutner*. Though she is only a little over two years old at the time of this writing, she is clearly one of the star children come onto our planet to help us all remember who we are. Thank you, Kendolynn. Thank you to all the children who are here now to guide and lead us into a wonderful future finding its way to Earth now. You are the light that will help us take the next steps.

And to the Twelve, whoever you are, wherever you are…Thank you for showing up and saying "yes." I trust our paths are already crossing.

BOOKS OF REFERENCE
AND EXCELLENT READS

Bohm, David):
- [and Mark **Edwards**] *Changing Consciousness: Exploring the Hidden Source of the Social, Political and Environmental Crises Facing our World (*1991)
- *Unfolding Meaning*: *A Weekend of Dialogue with David Bohm:* Donald Factor, Editor, Gloucestershire: Foundation House (1985)

Brown, Courtney:
- *Cosmic Voyage* (1996)

Card, Orson Scott
- *Songmaster* (1980)
- *Ender Series: Enders Game, Ender in Exile, Speaker for the Dead, Xenocide, Children of the Mind*

Carey, Ken:
- *The Starseed Transmissions* (1995)

Chopra, Deepak:
- *Creating Affluence (*1998)

Clark, Arthur
- *2001 Space Odyssey* (1968)

Eichler, Ruth:
- [and Lesley **Carmack**] *Radiant Awakening – Humanity's Transformational Journey*

Harman, Willis:
- *Global Mind Change (1998)*
- *Incomplete Guide to the Future (1979)*

Herbert, Frank:
- *Dune* (1965)

Hunter, John
- *World Peace and Other 4th Grade Achievements* (2013)

Jho, Zoev, Intergalactic Council Publications
- *ET 101: The Cosmic Instruction Manual* (1990)

Joynes, Monty
- *Journey for the One* (2008)

Kahane, Adam
- *Power and Love* (2010)

Kenyon, Tom
- *The Hathor Material* (2012)

Mack, John E., MD
- *Passport to the Cosmos* (1999)

MacLaine, Shirley
- *Sage-ing while Age-ing* (2007)

Marciniak, Barbara
- *Bringers of the Dawn* (1992)
- *Earth (1994)*

Maynard, Herman
- [and Sue **Mehrtens**] *The Fourth Wave* (1993)

Mehrtens, Sue
- *C.G. Jung and Our Collective future: Addressing the Challenges of Our Time through the Wisdom of Alchemy, Archetypes and Analytical Psychology* (2012)

McTaggart, Lynn
- *The Field* (2008)
- *The Intention Experiment* (2007)

Mitchell, Edgar
- *Psychic Exploration: A Challenge for Science, Understanding the Nature and Power of Consciousness* (1974)
- *The Way of the Explorer: an Apollo Astronaut's Journey through the Material and Mystical Worlds* (1996)

Mutwa, Credo
- *Song of the Stars* (2000)

Pehrson, John
- *Mystical Numerology: The Creative Power of Sounds and Numbers* (2013)
- [and Sue **Mehrtens**] *Intuitive Imagery: A Resource at Work* (2012)
- *Community Building: Renewing Spirit and Learning in Business* (1996) (participating author)

Plato
- *Plato's The Atlantis Dialogue* – Translated by B. Jowett, Shepherd Publications (2001)

Rael, Joseph
- *Being and Vibration* (1993)
- *Sound: Native Teachings and Visionary Art* (2009)

Renesch, John
- *Getting to the Better Future: a Matter of Conscious Choosing* (2000)
- *The Great Growing Up* (2012)

Sagan, Carl
- *Contact* (1985)

Sheldrake, Rupert
- *Morphic Resonance: the Nature of Formative Causation* (2009)

Sitchin, Zecharia
- *The 12th Planet* (1976)

White Eagle, Jeanne (Jeanne Lane **Pehrson**)
- *Grace – A Journey from Betrayal to Healing* (2012)
- *When the Canary Stops Singing: Women's Perspective on Transforming Business* (participating author as Jeanne Lane **Borei**)
- *Community Building: Renewing Spirit and Learning in Business*(1996) (participating author as Jeanne **Borei**)

Yogananda, Paramahansa
- *Autobiography of a Yogi* (1946, 2012)

Ywahoo, Dhyani
- *Voices of Our Ancestors* (1987)

Zukav, Gary
- *Dancing Wu Li Masters* (1984)
- *Seat of the Soul* (1990)

ABOUT THE AUTHOR

Jeanne White Eagle (Jeanne Lane Pehrson) is an educator and futurist with Cherokee roots. She is author of *Grace: A Journey from Betrayal to Healing* and co-author of two books on conflict resolution. She is a founding member of (Cast C) of the international musical "Up With People," and has worked in the business arena where she was awarded the prestigious Willis Harman Award by the World Business Academy. She is founder of *Up on the Mountain*, an organization for teens at risk and is co-founder of the For the One Organization whose purpose is to help remove walls of separation and promote global healing. She is an accomplished composer and professional classical singer.

Skilled in Conflict Resolution techniques, as of 1996 when she was given the name *Jeanne White Eagle* by Native American mystic, Joseph Rael, she has followed a vision that has taken her husband, John Pehrson, and herself around the world, often into war zones and other areas of deep conflict. With the use of Sound, Movement and the Open Heart Process, a process designed to help resolve conflict, she and her husband have become Ambassadors for worldwide peace. (In 2009, a biography of Jeanne and John was released, called *Journey for the One* by Monty Joynes.)

<>

Jeanne can be contacted at Jwhteeagle@aol.com. You're also invited to go to her website; www.Jeannewhiteeagle.com and Facebook page. At the time of this writing a new Blog is in the making.

<>

If you're interested in further information on John Pehrson and his system of Mystical Numerology go to www.mysticalnumerologyonline.com.

<><><>
<>

INDEX

13 - Month Calendar, 53

2001 Space Odyssey, 48, 183

4-sided pyramid, 82

4-sided square, 81

4th grade, 126, 127

Active Listening, 105, 132

acupressure points, 31, 79

Aether, 141

Agriculture Organization, 4

Ah, 89, 90, 91, 93

air, Air, 4, 5, 105, 106, 113, 134, 140, 141, 142, 143, 148

Air cars, 148

American Southwest, 23

angles, Angles, 105, 118

animal kingdom, Animal Kingdom, 106, 109, 121, 122

animals, 109, 122, 140

antahkarana, 116

Aqabal, 52

Arakis, 136

Ark, 75

astral, 106, 107

Atlantis, 106, 124, 125, 157, 158, 185

Atlantis Dialogues, 157, 185

Australia, 86

Autobiography of a Yogi, 186

Avatar, 159

awakening process, 122, 134

awareness, Awareness, 16, 19, 32, 35, 47, 57, 107, 114, 117, 120, 121, 128, 130, 132, 138, 139,

awareness training, 117

balance, Balance, 59, 106, 116, 122, 123, 130, 135, 139, 140

bankers, 127

bathrooms, Bathrooms (Accommodations for disposing, 105, 136

beacon, 32, 48, 56, 72, 83, 97, 120, 159, 173

Being and Vibration, 94, 186

Betazoid, 68

bi-polar magnet, 116, 174

bi-polar magnetic field, 116

blue magenta, 135

blueprint, Blueprint, 2, 32, 70, 83, 143

Bohm, David, 52, 97, 183

Breathable Air, 105, 134

breathing, Breathing, 39, 113, 132

Bridges, 55, 59, 83, 108, 119, 139

Bringers of the Dawn, 74-75, 184

Brown, Courtney, 71, 183

Buddhist, 40, 41

building blocks, 5, 140

Building Materials, 105, 118

Business, 8, 15, 23, 44, 185, 186, 189

Butterworth-Heinemann, 52

C.G. Jung, 52, 185

cafeterias, 135

Cafeterias, 105, 135

California, 70, 120, 121

cancer, 65, 91, 134

Carey, Ken, 102, 183

Carmack, Lesley, **184**

causality consciousness, 164

Center, 40, 41, 58, 59, 60, 61, 81, 82, 83, 89, 91, 97, 105, 108, 109, 115, 116, 124, 159, 164

Center Building, 58, 82, 83, 105, 109, 115, 116, 124

Center Building's Role in Energy Flow, 116

Center for Leadership Studies, 164

Center of Top Level, 105, 115

centeredness, 132, 135

central column, 41, 116

ceremonial dance, 15, 20

C.G. Jung and Our Collective future: Addressing the Challenges of Our Time through the Wisdom of Alchemy, Archetypes and Analytical Psychology, 185

Chamber, 95, 97

Changing Consciousness -Exploring the Hidden Source of the Social, Political and Environmental Crises Facing our World, 183

Changing Universe, 105, 128

chaos, 91, 125, 158

chasm of change, 165

Cherokee, 23, 40, 119, 189

Children, 14, 58, 66, 105, 106, 115, 124, 126, 138, 144, 146, 179, 183

Chopra, Deepak, 137, 183

Circles, 39, 40, 41, 51, 55, 82, 171, 172

civilization, *(see Foreword)*, 6, 7, 48, 157, 158

Close Encounters of the Third Kind, 66, 69

cold-blooded, 7, 112

collective consciousness, 52, 107, 117, 139, 140, 164

Colorado, 1, 91

colors, Colors 87, 105, 134, 135

Colors and Patterns, 105, 134

Communication, 3, 36, 48, 53, 61-69, 85, 96, 108, 111, 115, 118, 122, 123, 132, 133, 141, 145, 159

Community Building -Renewing Spirit and Learning in Business, 185, 186

compassion, Compassion, 153

conditioning areas, 111

conducting garments, Conducting Garments, 105, 130,136

conductive materials, 131

Conflict Resolution, 189, 190

Consciousness, *(see Foreword)*, 8, 20, 32, 43, 44, 47, 48, 52, 59, 65, 67, 90, 105, 107, 108, 109, 116, 117, 119, 121, 127, 128, 129, 130, 137, 139, 140-146, 153-157, 161, 164, 165, 183, 185

Contact, 69, 70, 186

Conversation Rooms and Active Listening, 105, 132

copper, 131

cosmic beacon, 83, 97

cosmic curiosity seekers, 73

Cosmic Voyage, 71, 183

Council, 2, 106, 144, 145, 146, 184

Creating Affluence, 137, 183

Croton-on-Hudson, 131

crystals, 56, 59, 60, 83, 93, 95, 96, 97, 98, 105, 115, 116, 118, 119, 122

Crystals, ii, 60, 93, 95, 97, 105, 119

Curriculum, 35, 85, 90, 97, 106, 108, 117, 125, 133, 139, 140, 144

Dalai Lama, 153

diplomats, 127

Directions, 40, 55, 89, 90, 91, 92, 94, 97

divine nature, 157, 158

divine portion, 157

DNA, 7, 74

Dune, 136, 184

early greeters, 115

Earth, *(see Foreword)*, 1, 3, 4, 5, 7, 21, 28, 32, 44, 61, 62, 66, 69, 71, 72, 73, 74, 79, 83, 88, 106, 109, 110, 111, 112, 113, 114, 116, 117, 119, 121, 122, 130, 131, 134, 137, 140, 141, 143, 145, 147, 162, 164, 180, 184

East, 40, 55, 56, 58, 59, 89, 90, 91, 97, 124, 127

education, 7, 31, 32, 46, 57, 163

Ee, 91, 93

Egyptian, 74

Eh, 91, 93

Eichler, Ruth, 184

Einstein, 43

Eisenhower, 15

elders, 58, 115, 124, 126, 144

electricity, 134

electromagnetic, 131

elementary exercises, 130

Elements, 106, 141, 142, 143

emotional, Emotional, 35, 40, 55, 56, 57, 90, 108, 125, 162

emotions, 57, 68, 109, 125, 130, 134

energy beacon, 120, 173

energy dance, 124, 133

energy flow, Energy Flow105, 116, 173

energy pyramid, 82, 83, 172

English, 23, 89

enhancers, 123

Enterprise, 136

environment, 36, 62, 68, 112, 113 118, 122, 128, 131, 132, 137, 155, 157, 158

ET 101, The Cosmic Instruction Manual, 1, 184

Ether, 141

evolution, Evolution, (*See Foreword*), 1, 3, 48, 72, 109, 120, 125, 140, 143, 145, 151, 156, 165

existence, 2, 4, 5, 62, 65, 67, 73, 79, 86, 87, 88, 116, 120, 143

Extrasensory Skills, 106, 124

Faraday cage, 131

Faraday room, 131, 132

Faraday Spaces and Silence, 131

Father of Africa, 4

feeling, 53, 85, 86, 114, 125, 154

feminine, 118

field of love, 155

Findhorn, 121

Finger of Fate, 18

Finger of God, 18

fire, Fire, 43, 121, 141, 142, 143

first dream, 147

Five Circles, 39, 40, 41, 51

Five Elements, 106, 141

Five Levels, 105, 108

food, Food, 4, 135, 136, 138

For the One Organization, 189

foundation, Foundation, 6, 8, 56, 60, 108, 110, 118, 158, 163, 183

Foundation House, 183

French, 23

frequency, 65, 98, 110, 111, 112, 117, 156

galaxies, 137, 159

Gardening, 106, 141

Gardens, 60, 90, 119, 120, 121, 128

Garments, 105, 130, 136

Geometries, 107

Germany, 39, 41, 51, 77, 79, 177

Getting to the Better Future - a Matter of Conscious Choosing, 7, 186

Global Mind Change, 5, 44, 184

Gloucestershire, 183

gold, 72, 81, 84

government, 163

Grace - A Journey from Betrayal to Healing, 186, 189

Grand Trines, 18

gravity, Gravity, 113, 114

Gray, 102

Greek, 74, 141

grounding, Grounding 112, 113

gut instinct, 142

Harman, Willis, 5, 43-46, 78, 180, 184, 189

Harvard, 8, 70

Hathor *Geometries*, 107

Hathors, 41, 102, 107, 142, 151, 153, 178, 184

heart, Heart, 45, 65, 106, 115, 124, 125, 148, 158, 190

heart energy, 115, 125

Hebrew, 51, 89

hemispherical, 115

Herbert, Frank, 136, 184

Hewlett Packard, 17

Hindu, 40, 41

holographic mindset, 107

Hook, 82

human beings, 3, 5, 28, 32, 47, 61, 98, 117, 122, 134, 137, 143, 156

human psyche, 122, 139, 144

humanity, Humanity, 7, 3, 4, 6, 63, 70, 73, 109, 153, 164

humankind, Humankind, 1, 4, 32, 47, 72, 78, 110, 111, 119, 120, 126, 130, 133, 139, 154, 155, 165

humans, Humans, 1, 4, 5, 7, 8, 11, 48, 62, 63, 64, 65, 66, 67, 68, 69, 72, 73, 77, 86, 89, 96, 105, 108, 109, 110, 112, 114, 115, 117, 119, 122, 124, 126, 129, 130, 131, 132, 133, 137, 139, 140, 142, 158

Hundredth Monkey Effect, 105, 129, 155

Hunter, John, 126, 127, 184

IBM, 17

ICC, 61, 62, 64, 67, 68, 70, 71, 72, 78, 79, 82, 83, 85, 87, 88, 89, 90, 93, 94, 97, 106-130, 133, 134, 137-149, 153-159, 171, 173, 174

Incomplete Guide to the Future, 44, 184

Indigenous peoples, 40, 56, 93, 121, 139

Indoctrination and Adaptation for Off-Planet Species, 105, 110-111

Institute of Noetic Sciences, (IONS), *(see Foreword)*, 44, 127

Intention, 67, 132, 185

interconnectedness, 120, 138, 145

Interdimensional, 108, 113, 114, 116, 117, 129

intergalactic, Intergalactic, 2, 62, 111, 116, 184

interstellar, Interstellar 28, 61, 73, 74, 75, 108, 126, 131, 159,

Interstellar Communication Center, 61, 159 (*see ICC and U/ICC)*

Intuitive, 52, 66, 68, 124, 126, 185

Intuitive Imagery - A Resource at Work, 52, 185

IONS, *(see Foreword)*, 44, 127

Irish, 23

Israel, 51, 87, 88, 153, 178

Japanese monkey, 129

Jeanne, 1, 3, 5, 7, 8, 2, 19, 23, 169, 186, 189, 190

Jeanne White Eagle, 1, 3, 5, 7, 8, 2, 23, 169, 189, 190

Jho, Zoev, 1, 184

JJ, 102, 178

Journey for the One by Monty Joynes, 5, 44, 91, 184, 190

Kenyon, Tom, 41, 102, 107, 142, 153, 184

land, 7, 4, 6, 106, 109, 140, 149

Language, 51, 64, 66, 68, 69, 70, 89

leadership, 164

Learning, 8, 31, 55, 105, 106, 108, 109, 117, 137, 138, 185, 186

Learning for Humans, 105, 109

library, 75

light, 30, 32, 34, 83, 92, 102, 105, 107, 112, 120, 134, 141, 147, 159,

lighting, Lighting, 105, 133, 147

linear, 17, 130, 137, 139

Listening, 87, 105, 132, 133

living organism, 145

Love, *(see Foreword)*, 36, 93, 105, 106, 109, 125, 144, 151, 153-157, 180, 184,

Lower Level and Conditioning Areas, 105, 111, 113

Macaca Fuscata, 129

Mack, John, 8, 156, 184

MacLaine, Shirley, 184

magic carpet, 128

magnetic field, 41, 116, 131, 174

Mandala, 40, 58

Manifesto, 35, 37, 43, 158

Marciniak, Barbara, 74, 184

mass consciousness, 47, 59, 108, 129, 140, 141, 145, 146, 156

Matrix, 73

Maynard, Herman, 179, 185

McTaggart, Lynn, 155, 185

Medicine Man, 23

meditation, Meditation, 43, 58, 60, 101, 113, 121, 139

meditation nooks, 139

Mehrtens, Sue, 17-20, 52, 179, 185

mental, Mental, 35, 40, 55, 56, 89, 90, 108, 124, 127, 162

metals, 118

military leaders, 127

Mission Control, 1

mission statement, 36

Mitchell, Edgar, *(See Foreword)*, 44, 52, 72, 127, 179, 185

molecular, 113

Monitoring Council, 106, 144, 145, 146

More on Children, 105, 106, 126

More on Children's Games, 126

morphic resonance, 155, 186

Morphic Resonance - The Nature of Formative Causation, 186

Morphogenetic Fields Theory, 52

moving walkways, 147, 148

music, 69, 86, 87, 88, 95, 139, 189

Mutwa, Credo, 4, 5, 185

Mystic, *(See Foreword)*, 23, 40, 92, 95

Mystical Numerology - The Creative Power of Sounds and Numbers, 52, 94, 185

NASA, *(See Foreword)*

National Public Radio, 126

Native American, 8, 23, 40, 51, 55, 56, 190

New York City, 131

Night Time Curriculum, 139

non-linear, 17, 139

Nordic, 23

North, 39, 40, 58, 59, 89, 91, 97

North Sea, 39

nourishment, Nourishment, 105, 135, 136

NPR, 126

Numbers, 52, 68, 69, 70, 94, 185

O, 89, 91

Off-Planet Species, 68, 105, 110

Oh, 89, 91, 93

one, One, 5, 44, 51, 91, 120, 180, 184, 190

Oneness, 120

OO, 89, 91, 93

Open Heart, 106, 124, 125, 190

Open Heart and Atlantis, 106, 124

oxygen, Oxygen, 46, 113

panda bears, 122

Paradigm, 127, 161, 164

parking lots, 148

Passport to the Cosmos, 8, 156, 184

patterns, Patterns, 96, 105, 134

peace dance, 20

Peace Sound Chamber, 95, 97

Pehrson, John, 52, 94, 180, 185, 190

Pentagon, 127

perineum, 41, 116

physical, Physical, 8, 16, 35, 40, 41, 48, 49, 56, 57, 61, 67, 71, 77, 79, 82, 85, 87, 90, 94, 106, 107, 108,

110, 112, 113, 116, 117, 121, 128, 130, 132, 134, 135, 136, 137, 139, 141, 142, 143, 156, 161, 162

physical body, 41, 136

physical Universe, 117, 128

Places for Nourishment (Cafeterias, etc.), 105, 135

Plant, 66, 106, 109, 120, 121, 122, 137, 140, 141

plant kingdom, Plant Kingdom, 66, 106, 120, 141

Plato, 157, 185

Pleiadian, 119, 178

politicians, 127

portal, 64, 98, 116, 149

prana, Prana, 41, 116, 174

prana tube, 41, 116, 174

primary buildings, 55, 56, 60, 90, 108

primary vowel sounds, 89

principle ideas, 89

Programmed Sounds, 116, 117

psionic, 68

psychiatry, 131

psychic, Psychic, 66, 67, 68, 126, 185

Psychic Exploration - A Challenge for Science, Understanding the Nature and Power of Consciousness, 185

purification system, 134

Pyramid, 72, 81 82, 83, 97, 107, 115, 172, 173

Qi, 41

rabbi, 153, 154

Radiant Awakening - Humanity's Transformational Journey, 184

radiation, 131

radio, Radio, 70, 126

Rael, Joseph, 40, 92, 94, 95, 180, 186, 190

reality, Reality, 5, 17, 25, 35, 44, 48, 71, 77, 86, 87, 90, 94, 96, 107, 113, 137, 139, 161, 162

Recurring Dream, 3, 6, 13, 16, 20, 24, 33, 39, 60, 61, 77, 78, 83, 101, 106, 114, 124, 125, 129, 147, 148, 153, 155, 156, 158, 164, 169, 170, 177

Relationship to Water, Land, Air, 140

Renesch, John, 163, 164, 178, 186

Roman, 74

Russian cosmonaut, 74

Sagan, Carl, 186

Sage-ing while Age-ing, 184

San Francisco, 127

San Juan Mountains, 91

Sangoma, 4

Sanusi, 4

science fiction, *(See Foreword)*, 14, 60, 63, 68, 71, 73, 147, 161

Scotland, 121, 178

Scottish, 23

sea, Sea, 5, 27, 39, 129, 140

Search for Extraterrestrial Intelligence, 70

Seat of the Soul, 187

second dream, 92, 147

sentient, 97, 119

separation, 20, 162, 189

Separation, 162

SETI, 70

shape, Shape, 27, 49, 55, 56, 71, 81, 82, 83, 87, 95, 107, 115, 116, 157

Sheldrake, Rupert, 52, 197, 155, 186

shift in consciousness, 161

signals, Signals, 48, 70, 72, 83, 102, 173

Silence, 105, 131, 132

Simultaneous Learning, 106, 137

Sitchin, Zecharia, 186

solar, 133

song, Song, 86, 88, 93, 95, 102, 128, 138, 183, 185

Song of the Stars, 185

Songmaster, 183

Sound, 40, 55, 60, 65, 69, 85-97,, 105, 116, 117, 128, 133, 134, 139, 142, 185, 186, 190

Sound - Native Teachings and Visionary Art, 186

Sound Communication, 65

sound sensitive devices, 134

Source, 41, 58, 108, 154, 183

South, 40, 55, 56, 57, 59, 89, 90, 91, 97

South Africa, 5

space explorers, 73

space-time portals, 64, 98, 116, 149

Spanish, 89

spatial coordinates, 114

spatial perception, 138

Spielberg, 69

spiritual, Spiritual, 8, 35, 40, 58, 108, 162

Spontaneous Sound, 65, 90

squares, 81, 82, 84

Star Beings, 36, 62, 75

Star Trek, 6, 7, 68, 73, 136

Star Trek -The Next Generation, 68

Star Trek Voyager, 7

Star Wars, 99, 136

Stargate, 73

Starseed Transmissions, 102, 183

subspace, Subspace, 71, 106, 113, 117, 136, 139

subspace beings, 71, 113

symbiotic relationship, 79, 140

sympathetic vibrations, 96

telepathic, 66, 67, 137

telepathy, 67, 124

Tennessee, 18, 120, 121

Texas, 44

The 12th Planet, 186

The Dancing Wu Li Masters, 52, 187

The Field, 138, 155, 185

The Five Elements, 106, 141

The Fourth Wave, 185

The Great Growing Up, 163, 186

The Hathor Material, 41, 102, 107, 151, 153, 184

The Intention Experiment, 185

The Mind's Eye, 11

Thirteen, 51, 52, 53, 55, 82, 144, 171, 172

Thirteen Circles, 55, 82, 171, 172

thought, 56, 65, 86, 88, 91, 97, 114, 129, 134, 138

three-dimensional, 81, 143

threshold, 164, 165

top level, 105, 109, 115

Transit System, 105, 106, 124, 147-149

Transmissions, 41, 60, 93, 98, 99, 101, 102, 105, 144, 147, 148, 155, 183

transparent tubes, 148

Transportation System, 124

tree, 86, 96, 120, 121

tribal leaders, 127

Tube, 41, 116, 148, 174

tube torus, 116

Tunnels, 106, 123

Twelve, 13, 24, 79, 164, 181

Twenty-Four Hour Curriculum, 106, 139

Two-dimensional, 174

149, 153, 154, 155, 156, 157, 158, 159, 171, 173, 174

U/ICC, 61, 67, 68, 70, 71, 72, 78, 79, 82, 83, 85, 87, 88, 89, 90, 93, 94, 97, 106, 107, 108, 109, 110, 111, 112, 113, 114, 115, 116, 117, 118, 119, 120, 122, 123, 124, 125, 126, 127, 128, 130, 133, 134, 137, 138, 139, 140, 142, 143, 144, 145, 147, 148, 149, 153, 154, 155, 156, 157, 158, 159, 171, 173, 174

underwater, Underwater 105, 109

underground, Underground 105, 109, 111, 112, 115, 123, 147

Underground and Underwater Effect, 105, 109

Unfolding Meaning - A Weekend of Dialogue with David Bohm, 183

United Nations, 4, 92

United Nations Food and Agriculture Organization, 4

Universe, 2, 5, 35, 39, 41, 49, 77, 78, 79, 105, 109, 117, 123, 128, 137, 140, 148, 154, 164

Universe Cities, 77, 79, 109, 140

Universe City, 78, 123, 148

University, 31, 35, 39, 46, 48, 49, 53, 55, 56, 57, 58, 59, 60, 61, 67, 70, 77, 78, 125, 131, 159, 164

Up With People, 15, 180, 189

utopia, 158

Vibration, 51, 55, 65, 67, 68, 85-96, 102, 105, 106, 110, 111, 113, 114, 117, 119, 128, 130, 132, 133, 134, 135, 144, 153-157, 186

Vibration Increase, 105, 130

Vibration of Love, 106, 144, 156

vision, Vision, 15, 16, 20, 33, 41, 43, 46, 48, 49, 61, 71, 77, 91, 95, 159, 161, 177, 178, 190

Visitors, 28, 36, 61, 62, 63, 64, 66, 67, 69, 71, 72, 73, 74, 110, 111, 112, 113, 114, 115, 117, 120, 126, 130, 131, 132, 134, 136, 156

Volunteers, 79

vowels, 89

Vusamazulu, 4

warm-blooded, 7, 112

Wasserkoog, 39

waste material, 105, 136

water, Water, 4, 105, 106, 109, 111, 120, 122, 123, 129, 136, 140, 141, 142, 143, 147

Way of the Explorer, 52, 72, 185

West, 40, 57, 59, 89, 91, 97

Western, 23, 67

When the Canary Stops Singing:, 186

Willis Harman Award, 189

Willis Talks, 43, 45, 46, 78

wordless songs, 128

World Business Academy, 15, 44, 189

World Peace and other 4th Grade Achievements, 127, 184

Wright Brothers, 11

writing lab, 101, 179

Yale University, 131, 164

Yale University School of Medicine, 131

Yods, 18, 19

Yogananda, Paramahansa, 68, 186

Ywahoo, Dhyani, 119, 186

Zoev Jho, 1, 184

Zukav, Gary, 187

Made in the USA
Middletown, DE
21 December 2021